Magneto-Resistive Heads
FUNDAMENTALS AND APPLICATIONS

ELECTROMAGNETISM

Electromagnetism is a classical area of physics and engineering which still plays a very important role in the development of new technology. Electromagnetism often serves as a link between electrical engineers, material scientists, and applied physicists. This series will present volumes on those aspects of applied and theoretical electromagnetism that are becoming increasingly important in modern and rapidly developing technology. Its objective is to meet the needs of researchers, students, and practicing engineers.

This is the first volume in the series. It deals with the fundamentals of magneto-resistive heads, an area of growing importance in the magnetic recording industry. It is my hope that this volume will be well received by the international scientific and engineering community involved in research and development in the field of magnetic recording.

ISAAK MAYERGOYZ
DEPARTMENT OF ELECTRICAL ENGINEERING
UNIVERSITY OF MARYLAND

This is a volume in

ELECTROMAGNETISM

ISAAK MAYERGOYZ, SERIES EDITOR
UNIVERSITY OF MARYLAND, COLLEGE PARK, MARYLAND

A list of titles in this series appears at the end of this volume.

Magneto-Resistive Heads

FUNDAMENTALS AND APPLICATIONS

JOHN C. MALLINSON

Mallinson Magnetics, Inc.

Belmont, California

ACADEMIC PRESS

San Diego New York Boston London

Sydney Tokyo Toronto

Cover photograph courtesy of the IBM Corporation
Research Division, Almaden Research Center.

This book is printed on acid-free paper. 

Academic Press, Inc.
A Division of Harcourt Brace & Company
525 B Street, Suite 1900, San Diego, California 92101-4495

United Kingdom Edition published by
Academic Press Limited
24-28 Oval Road, London NW1 7DX

Library of Congress Cataloging-in-Publication Data

Mallinson, John C.
Magneto-resistive heads : fundamentals and applications / by John C. Mallinson.
p. cm. -- (Electromagnetism series)
Includes bibliographical references and index.
ISBN 0-12-466630-2 (alk. paper)
1. Magneto recorders and recording--Heads. I. Title.
II. Series.
TK5984.M35 1995
621.382'34--dc20 95-16539
CIP

PRINTED IN THE UNITED STATES OF AMERICA
95 96 97 98 99 00 QW 9 8 7 6 5 4 3 2 1

This book is dedicated to my wife, Phebe,
who provided the principal inspiration for
my undertaking the task of writing this book.

Contents

CHAPTER 3

The Reading Process

CHAPTER 4

The Anisotropic Magneto-Resistive Effect

CHAPTER 5

Vertical Biasing Techniques

CHAPTER 6

Horizontal Biasing Techniques

CHAPTER 12

The Giant Magneto-Resistive Effect

CHAPTER 13

Spin Valve and Granular Giant Magneto-Resistive Heads

CHAPTER 14

Simplified Design of a Shielded Magneto-Resistive Head

CHAPTER 15

Read Amplifiers and Signal-to-Noise Ratios

Appendix

Preface

In 1968, the Ampex Corporation Research Department hired a recently graduated doctoral student from the Massachusetts Institute of Technology named Robert P. Hunt. His initial assignment was quite simple: "Find something new and useful in magnetic storage technology." Shortly thereafter, Bob Hunt invented the magneto-resistive head (MRH).

I had the privilege of sharing an office with Bob Hunt and not only witnessed his invention but also provided the first analysis of the output voltage spectrum for both horizontal and vertical unshielded magneto-resistive heads. I have followed MR technology closely in the ensuing 27 years and this book is the result.

Of greater importance is the fact that the office immediately adjacent to mine at Ampex Research was occupied by Irving Wolf, the inventor of the "Wolf" permalloy electroplating bath. When asked by Bob Hunt if he could fabricate a magneto-resistive head, Irving Wolf replied, "Certainly, I'll make it out of evaporated permalloy." This was, indeed, a remarkable piece of serendipity. It is quite possible that, had Bob Hunt and Irv Wolf not been in such close proximity, there would be no MRHs even today. Not only did the first MRHs use permalloy as the sensor, but all MRHs manufactured in large quantities to this day also use permalloy. Moreover, it is likely that most of the advanced MRH designs now being proposed will also use permalloy.

The intent of this book is to introduce the reader to the principal developments in MRH technology that have occurred since Bob Hunt's initial work. To make this book self-contained, Chapter 1 contains a review of the basics of magnetic materials and magnetism and Chapters 2 and 3 cover the writing process and the usual inductive reading process in magnetic recording, respectively. The anisotropic magneto-resistive effect and the unique properties of permalloy are discussed in Chapter 4.

To achieve linear operation with low harmonic distortion, the MR sensor must be magnetically biased with a vertical bias field. Additionally, a horizontal bias field is usually applied to keep the magnetic state of the MR sensor stable. The principal techniques for producing these biasing fields are the subjects of Chapters 5 and 6.

In Chapter 7, the output voltage spectra and isolated written transition output pulse shapes of Hunt's horizontal and vertical MRHs are reviewed. A similar analysis of the very commonly used "shielded" MRH follows in Chapter 8. Here, the important fact is demonstrated that the output voltage spectral shape and isolated pulse shape of the shielded MRH and an ordinary inductive read head are almost identical.

Alternative designs for MRHs in which the MR element is incorporated into the structure of thin-film ring heads are considered in Chapter 9. These types of MRHs are often called "yoke-type" or "flux-guide" designs.

Considerable interest exists in the performance characteristics of MRHs which have two MR sensors. Depending on the directions of magnetization, the current, and the external voltage sensing connections to these double-element heads, they are called a variety of rather confusing names such as gradiometer, dual-stripe, and dual-magneto-resistive heads. These differences and their relative performance advantages and disadvantages are the topic of Chapter 10.

In Chapter 11, the output voltages of a single-element shielded MRH and an inductive read head are studied in ways that lead to a particularly simple and direct way of comparing their performance.

An entirely new physical phenomenon called the giant magneto-resistive (GMR) effect was discovered in 1987. In Chapter 12, the basic physics of this phenomenon is outlined and, in Chapter 13, a proposed design for some giant magneto-resistive heads (GMRHs) is discussed. It is expected that GMRHs will produce output voltages greater by perhaps a factor of 5 than those of conventional anisotropic MRHs.

The penultimate chapter contains a step-by-step simplified design sequence for a single-element shielded MRH. This material is included because it demonstrates in a straightforward fashion precisely which physical phenomenon controls each of the principal dimensions of the MR sensor and the MR sensor-shield "half-gaps."

The last chapter is devoted to system considerations such as read amplifier designs and the signal-to-noise ratio of MRHs. Finally, conclusions concerning the significance of MRHs in future high-density digital recorders are offered.

An appendix which contains a listing of the defining equations and a table of conversion factor for cgs–emu and MKS-SI magnetic units is provided. Despite the fact that virtually all physicists and electrical engineers in the world are trained to use MKS-SI, the magnetic storage industries in the United States and Japan continue to use cgs–emu, and this convention is followed in this book. In my opinion, the preference for one convention over another is an unimportant matter of taste.

Most of the material in this book is presented in a nonmathematical manner. Thus there are no mathematical analyses or derivations. On the other hand, the results of such analyses and derivations are used extensively. My belief is that most readers are not interested in detailed mathematical formalism and often find that copious mathematical detail often acts to obscure the intuitive obviousness of the underlying physics. The lack of mathematics should not, however, lead the reader to suppose that this is an elementary exposition. On the contrary, the aim is to lead the reader, in a straightforward fashion, to a high level of scientific understanding of MR materials and heads.

Finally, my hope is that this book will prove enjoyable and fascinating to read. Even after spending the last 35 years in magnetic recording research and theory, I find the seemingly unending inventiveness of my industrial colleagues both amazing and glorious. I hope that this book helps others to share in this rich and fascinating field.

John C. Mallinson
Belmont, California

Acknowledgments

The author of this book acknowledges the help of his many friends in the magnetic recording industry. Karen Bryan produced all of the figures and Phebe Mallinson did the word processing.

CHAPTER

1

B, H, and M Fields

To understand magnetic recording and magneto-resistive heads (MRHs), we must first distinguish between the three fields B, H, and M. Each is a field which, at all points in three-dimensional space, defines the magnitude and direction of a vector quantity. The field B is called the magnetic flux density, H the magnetic field, and M the magnetization. All have properties similar to those of other, perhaps more familiar, fields such as the water flow in a river, the airflow over a wing, or the earth's gravitation.

Some arbitrariness exists concerning the order in which B, H, and M are introduced. Here the magnetic field is defined first, because it can be related to simple classical experiments. Magnetism and magnetization are treated next as logical extensions of the concept of a magnetic field. Finally, the magnetic flux density is discussed as the vector sum of the magnetic field and the magnetization.

Magnetic Field H and Magnetic Moment μ

When an electric current flows in a conductor, it produces a magnetic field in the surrounding space. Suppose that current I flows in a small length, dl, of the conductor. The magnetic field, dH, measured at a distance R from the conductor, is orthogonal to both the current flow direction and the measuring distance vector. The magnitude of the magnetic field is given by the inverse square law and is

$$dH = 0.1 I dl / R^2, \tag{1.1}$$

where H is in oersteds, I is in amperes, and both l and R are in centimeters. All magnetic quantities in this book are given in the centimeter, gram, second, electromagnetic (cgs-emu) system of units. Conversion factors to the Système Internationale of metric units (SI) are given in the appendix.

Now consider a very long, straight conductor as shown in Fig. 1.1. By integrating Eq. (1.1), we can see that the magnetic field produced circles the conductor. The magnetic field is everywhere tangential, that is, orthogonal to both the current direction and the radius vector r. The direction of the magnetic field is given by the right-hand rule. Point the thumb of the right hand in the direction of the current flow and the magnetic field direction is given by the way the curved fingers point. The magnitude of the field is given by

$$H = 0.2I/r, \tag{1.2}$$

where r is the radial distance in centimeters.

Now suppose that the conductor is coiled to form the solenoid shown in Fig. 1.2. The magnetic field inside a long solenoid is nearly uniform (parallel and of the same magnitude) and is

$$H = 0.4\pi NI/1, \tag{1.3}$$

where N is the number of turns and l is the solenoid length in centimeters. Note that the field does not depend on the cross-sectional area of a long solenoid. At this point, it is convenient to define another vector quantity, the magnetic moment,

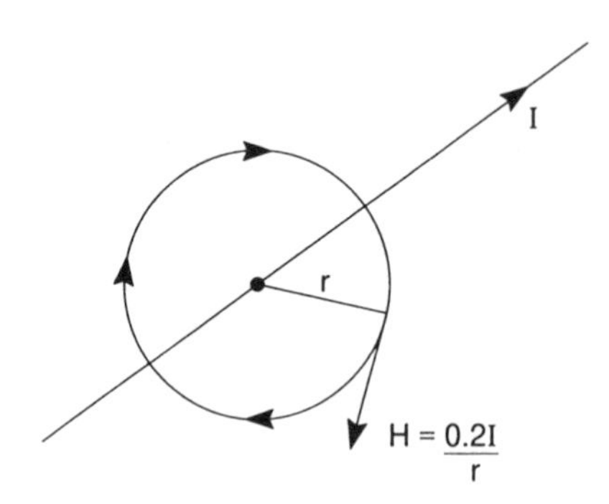

Fig. 1.1. The tangential field of a long, thin conductor.

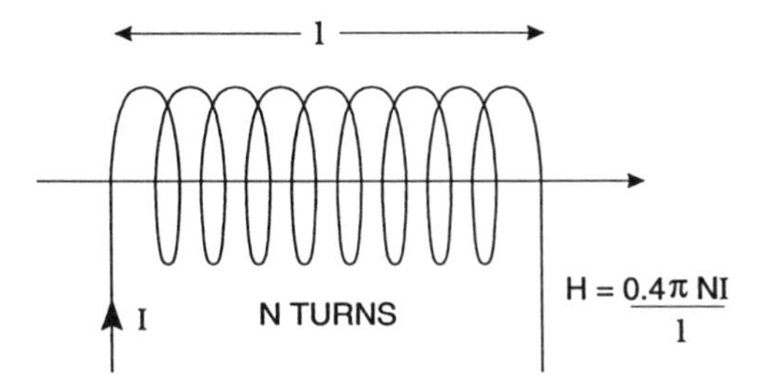

Fig. 1.2. The axial field of a long solenoid.

$$\mu = \frac{1}{4\pi} \int H dV. \tag{1.4}$$

Here, μ is the magnetic moment in electromagnetic units (emu) and V is volume in cubic centimeters. For any solenoid, it can be shown that

$$\mu = 0.1NIA, \tag{1.5}$$

where A is the vector normal to the area of the solenoid cross section, whose magnitude is the solenoid cross-sectional area in square centimeters.

Electron Spin

Magnetization is a property which arises from the motions of electrons within atoms. The magnetization of *free space,* that is, space free of material bodies, is by definition zero.

In all atoms, electrons orbit a nucleus made up of protons and neutrons. The atomic number is the number of protons which, for an electrically neutral (not ionized) atom, equals the number of electrons. The atomic weight is the number of protons plus the number of neutrons. Within an atom, the electron has two separate motions. First, the electron orbits the nucleus, much as the earth orbits the sun, at a radius as small as a few angstroms (10^{-8} cm). Second, the electron spins on its own axis, much as does the earth daily. These two motions are, of course, governed by the laws of quantum physics. For isolated atoms, the laws are known as Hund's rules. All motions of an electron produce an electric current and, just as the motion of electrons around a solenoid produces a magnetic moment, the orbital motion gives rise to an orbital magnetic moment and the spinning motion causes an electron spin magnetic moment. Generally, Hund's rules prescribe that the several electrons in an atom orbit in opposite directions, so that the total orbital moment is small. In the solid state, interactions with neighboring atoms "quench," that is, reduce further, the orbital moment. In the materials used in magnetic recording and magneto-resistive heads, the first transition group of elements, the contribution of the orbital magnetic moments to the magnetization is virtually negligible.

The magnetic moment of a spinning electron is called the *Bohr magneton* and is of magnitude

$$\mu_B = \frac{eh}{4\pi m} = 0.93 \times 10^{-20} \text{ emu} \tag{1.6}$$

where e is the electron's charge in electromagnetic units (1.6×10^{-20}), h is Planck's constant (6.6×10^{-27}), and m is the electron's mass in grams (9×10^{-28}).

The spinning electron has a quantum spin number, $s = \frac{1}{2}$, and it can be oriented (in a weak magnetic reference field) in only $(2s + 1) = 2$ directions. For atoms in free space, Hund's rules normally ensure that the electrons' spin directions alternate, so that the total electron spin moment is no more than one Bohr magneton. Moreover, this small moment is reduced in the solid state by next-neighbor interactions. Fortunately, however, nature has provided irregularities in the electron spin ordering. In transition group elements, outer electron orbits, or shells, begin to be occupied before the inner ones are completely filled. The result of this is that the inner partially filled shells can have large net electron spin moments and yet have the neighboring atom's interactions be partially screened off by the outer electrons.

Consider an atom of iron in free space, as depicted in Fig. 1.3. The atomic number of iron is 26, and there are, accordingly, 26 electrons. The electron shells are shown numbered by two quantum numbers. The first, or principal, quantum number (n = 1, 2, 3, and 4) is related to the electron's energy. The orbital quantum number (s, p, and d) defines the orbital shape. Note that in shells 1s, 2s, 2p, 3s, 3p, and 4s, equal numbers of electron spins point up and down so that the total electron spin moment is zero; that is, the electron spins are "compensated."

In shell 3d, however, an uncompensated spin moment of 4 μ_B exists because there are five spins up but only one down. Higher in the periodic

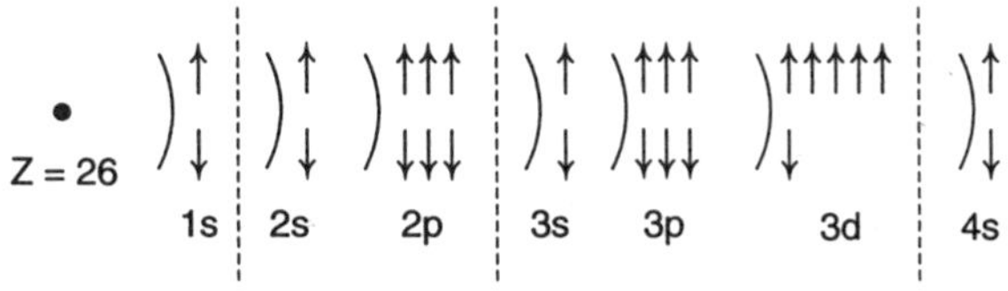

Fig. 1.3. The electron spin orientations of an iron atom in free space.

table, the $3d$ shell eventually fills, resulting in 10 electrons with five up and five down and zero net electron spin moment. The first transition group elements (Cr, Mn, Fe, Co, and Ni), however, have $3d$ shells unfilled and have uncompensated electron spin magnetic moments. Nearly all practical interest in magnetism centers on the first and second transition groups of elements with uncompensated spins.

When iron atoms condense to form a solid-state metallic crystal, the electronic distribution, called the *density of states,* changes. Whereas the isolated atom has $3d$; 5+, 1−, $4s$; 1+, 1−, in the solid state the distribution becomes $3d$; 4.8+, 2.6−, $4s$; 0.3+, 0.3−. Note that the total number of electrons remains 8, but that the uncompensated spin moment is lowered to 2.2 μ_B. The screening of the neighboring atoms by the $4s$ electrons is imperfect. For all practical purposes, an iron atom in metallic alloys has a 2.2 μ_B of magnetic moment.

Exchange Coupling

Now consider the magnetic behavior of iron atoms in an iron crystal. The crystal form, called the *habit,* is body-centered cubic with a cube-edge dimension of 2.86 Å. The first question to ask concerns the relationship of the atomic moment of one iron atom to that of its neighbors. A quantum effect, called *exchange coupling,* forces all the iron atom's magnetic moments to point in nearly the same direction. Exchange coupling lowers the system's energy by aligning the uncompensated moments.

At absolute zero, the ordering is perfect, whereas at higher temperatures, thermal energy causes increasing disorder. At the Curie temperature, in iron 780°C, thermal energy equals the exchange energy and all long-range order breaks down. Thus, the spin moments point randomly in all directions. In this disordered state, the material is said to be a *paramagnet* (that is, almost a magnet). Below the Curie temperature, the parallel alignment is called *ferromagnetism.*

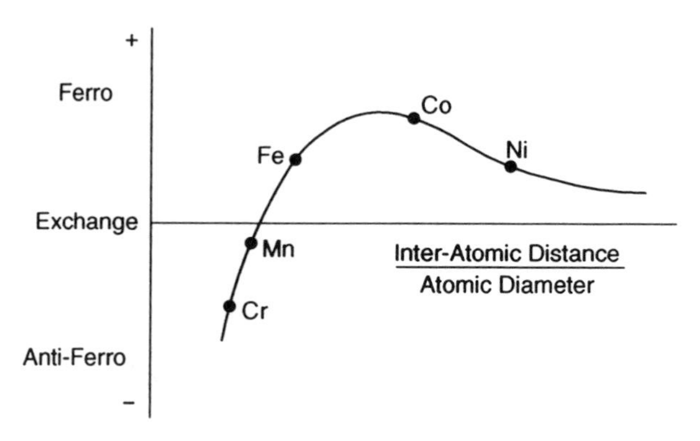

Fig. 1.4. The exchange coupling between magnetic atoms as a function of their spacing-to-size ratio.

Other orderings of the atomic moments are found in nature. Exchange coupling between atoms depends sensitively on the ratio of the interatomic distance to the atomic size as is indicated in Fig. 1.4. When the atoms are relatively closely spaced, as in the cases of Cr and Mn, the exchange coupling is negative and the adjacent spins are aligned in an antiparallel manner. Such materials are referred to as *antiferromagnets* and, of course, they have zero magnetic moment.

For larger spacing ratios, the exchange coupling is positive and the spins are aligned parallel as in the ferromagnetic metals Fe, Co, and Ni. In yet other magnetic materials, for example, γ-Fe_2O_3 and the ferrites, an intermediate ordering called *ferrimagnetism* exists, where the number of spins in each direction is unequal. These three spin orderings are shown in Fig. 1.5.

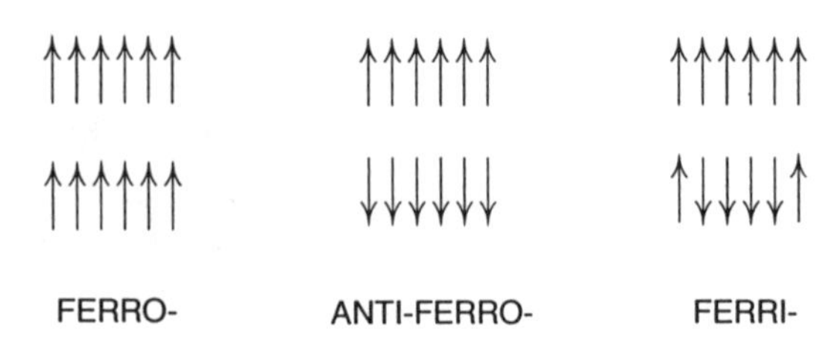

Fig. 1.5. Three common types of magnetic ordering.

Magneto-Crystalline Anisotropy *K*

Another question to ask about iron in the solid state concerns the orientation of the net magnetic moment with respect to the crystal axes. It turns out that the ferromagnetic-ordered atomic moments are aligned parallel to the body-centered cube edges. In iron, the cube edges are the easy, or lowest energy, directions, with the body diagonals being the hard, or highest energy, directions of the magnetic moment. A measure of this energy difference is the magneto-crystalline anisotropy constant, K. It is the energy required, in ergs per cubic centimeter, to rotate the magnetic moments from the easy to the hard direction. Several different symmetries of magneto-crystalline anisotropy are found in nature. In magnetic recording technology, most interest centers on the cubic, as in iron and γ-Fe_2O_3, and the uniaxial, as in iron-nickel (for example, permalloy) alloys.

Magnetization *M*

Now let our perspective expand to include a volume of iron that contains several million atoms. Just as previously our viewpoint enlarged from the electron spin level to the atomic level, now the focus is on millions of atoms. The magnetization is, by definition, the volume average of the atomic moments:

$$M = \frac{1}{V} \sum_{1}^{N} m \tag{1.7}$$

where V is the volume in cubic centimeters, m is the atomic moment in emu, and N is the number of atomic moments in the volume V. The units of magnetization are, therefore, magnetic moment per unit volume or electromagnetic units per cubic centimeter (emu/cm^3). In a large enough magnetic field, the magnetization of all parts of a magnetic material is parallel; at lower fields the magnetization may subdivide into domains. It is to be noted that within a single domain the magnetization is parallel everywhere and uniform and has a value called the saturation magnetization, M_s. The value of M_s depends on the temperature, being a maximum at absolute zero and vanishing at the Curie temperature.

The 0°K value of M_s for a body-centered cubic iron crystal can be calculated easily. Each iron atom has 2.2 μ_B of magnetic moment; there are, on average, two iron atoms per unit cell; and the cell edges measure 2.86 Å. It follows that

$$M_s\ (T = 0) = \frac{2.2 \cdot \mu_B \cdot 2}{(2.86 \times 10^{-8})^3} = 1700\ \text{emu/cm}^3 \tag{1.8}$$

At room temperature, M_s is only slightly reduced by thermal energy, so that for all practical purposes, pure iron has the following properties: $M_s = 1700$ emu/cm³, $4\pi M_s = 21{,}000$ G, and $\sigma_s = 216$ emu/g, where σ_s is called the specific saturation magnetization.

The values of $4\pi M_s$ for some other materials of interest in magnetic recording are cobalt, 18,000 G; nickel, 6000 G; and permalloy (81 Ni/19 Fe), 12,500 G.

Magnetic Poles and Demagnetizing Fields

In general, when a magnetic material becomes magnetized by the application of a magnetic field, it reacts by generating, within its volume, an opposing field that resists further increases in the magnetization. This opposing field is called the *demagnetization field* because it tends to reduce or decrease the magnetization. To compute the demagnetization fields, first the magnetization at all points must be known. Then, at all points within the sample, the magnetic pole density is calculated,

$$\rho = -\nabla M = -\left(\frac{dM_x}{dx} + \frac{dM_y}{dy} + \frac{dM_z}{dz}\right), \tag{1.9}$$

where ρ is the pole density (emu/cm⁴), and M_x, M_y, and M_z are the orthogonal components of the magnetization vector. The convention for magnetic poles is that when the magnetization decreases, the poles produced are north or positive. It is an unfortunate historical accident that the earth's geographic north pole has south, or negative, magnetic polarity.

Magnetic poles are of extreme importance because they also generate magnetic fields, H. The only two sources of magnetic fields are *real* electric currents and magnetic poles. The adjective *real* is used to distinguish real currents flowing in wires, which may be measured with ammeters, from hypothetical currents flowing in atoms due to their orbiting and spinning electrons. The magnetic fields caused by magnetic poles can be computed using the inverse square law. The field points radially out from the positive, or north pole, and has the magnitude,

$$dH = 4\pi\rho dV/r^2, \tag{1.10}$$

where H is the magnetic field in oersteds, ρ is the pole density, V is the volume (cm^3), and r is the radial distance (cm). Note that magnetic poles are analogous to electric poles.

Most field computations in magnetic recording are two dimensional. This is because one dimension, the track width, is very large compared with the other two. Two-dimensional magnetic fields have many simplifying properties, and one of the most important of these is shown as Fig. 1.6. The magnetic field from a long straight line of poles points radially out and has magnitude

$$H = 0.2s/r, \tag{1.11}$$

where H is the magnetic field in oersteds, s is the pole strength per unit length (emu/cm^2), and r is the radial distance in centimeters. The crucial thing to notice is that Eqs. (1.2) and (1.11) have the same form. Apart from scaling factors, a change from electric current to magnetic poles causes a 90° rotation of the magnetic field at every point in a two-dimensional space. The magnetic fields from currents and poles are orthogonal.

In general, the fields generated by a magnetic body are very complicated and force the magnetization to be nonuniform. For one class of geometrical shapes, however, it is known that the demagnetizing field and magnetization can be uniform. When any ellipsoid is uniformly magnetized, the demagnetizing field is also uniform. The demagnetizing field can be written

$$H_d = -NM, \tag{1.12}$$

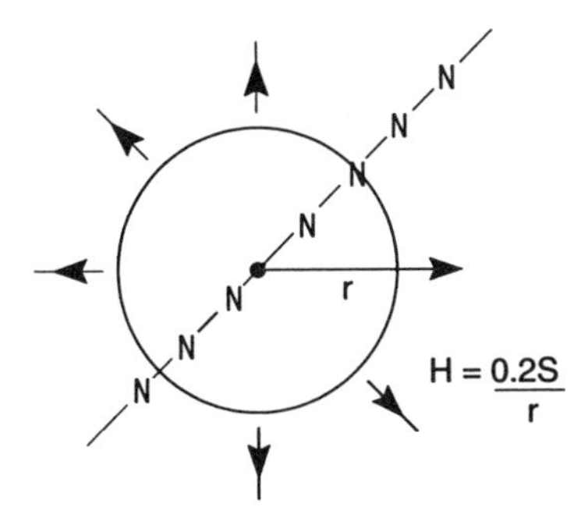

Fig. 1.6. The radial field from a long line of poles.

where H_d is the vector demagnetizing field, N is the demagnetization tensor, and M is the vector magnetization. For ellipsoids, the demagnetization tensor is the same at all points within a given body. Note that since "tensors turn vectors," the demagnetizing field need not be exactly antiparallel to the magnetization. Ellipsoids of revolution range from infinite flat plates through oblate spheroids to spheres, through prolate spheroids to infinite cylinders. They are formed by rotating an ellipsoid about either its major or minor axis. The demagnetizing tensors for three cases are shown below:

$$\begin{array}{ccc} xx & xy & xz \\ yx & yy & yz \\ zx & xy & zz \end{array} \quad \begin{array}{ccc} 0 & 0 & 0 \\ 0 & 0 & 0 \\ 0 & 0 & 4\pi \end{array} \quad \begin{array}{ccc} 4\pi/3 & 0 & 0 \\ 0 & 4\pi/3 & 0 \\ 0 & 0 & 4\pi/3 \end{array} \quad \begin{array}{ccc} 2\pi & 0 & 0 \\ 0 & 2\pi & 0 \\ 0 & 0 & 0 \end{array}$$

Thus, the flat plate has no demagnetization within its x-y plane but suffers a 4π demagnetizing factor on magnetization components out of the plane. A sphere suffers a $4\pi/3$ factor in all directions. A long cylinder has no demagnetization along its axis, but suffers 2π in the x and y directions of its cross sections. Note that these tensors are all diagonal, because the axis of rotation coincides with the z direction, and that the diagonal terms always sum to 4π. This is because 4π steradians of solid angle fill three-dimensional space. In cgs-emu, 4π field lines emanate from a unit magnetic pole. In other systems of units (for example, MKS-SI) the 4π appears in other places, but its appearance cannot, of course, be suppressed entirely.

Consider an ellipsoidal sample of magnetic material within a long solenoid that produces an axial field, H_s, as shown in Fig. 1.7. Suppose that the solenoid field is strong enough to saturate the magnetization. The magnetization of the ellipsoid is uniform, and thus there are no magnetic poles within the volume. Poles form, however, on the surfaces as shown by the letters N and S in the figure. These surface poles produce a demagnetizing field, H_d, which is exactly antiparallel to both the magnetization and the solenoid field. If the ellipsoid axis had not been parallel to the solenoid axis, these exact alignments would not have occurred. It is clear now that the total magnetic field, H_t, within the sample is

$$H_t = H_s - H_d. \tag{1.13}$$

The effect of the demagnetizing process is to reduce the field inside the sample in a manner which is exactly analogous to that of the negative feedback system.

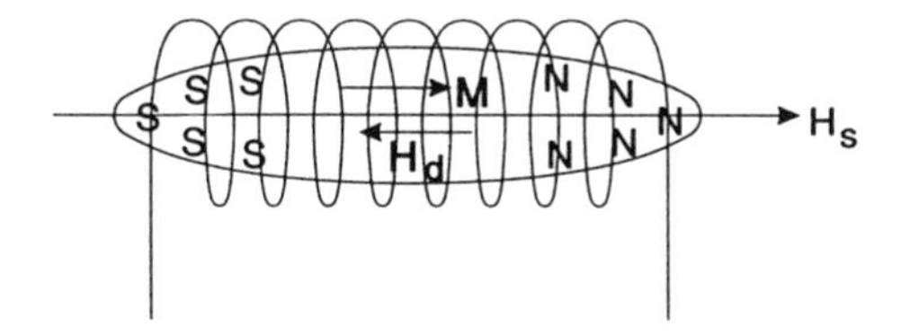

Fig. 1.7. An ellipsoid of revolution in a solenoid, showing the induced magnetic poles and the demagnetizing field.

Flux Density B and Flux ϕ

Having defined and discussed both the magnetic field H and the magnetization M, the flux density B can now be defined as

$$B = H + 4\pi M, \tag{1.14}$$

where B is the flux density in gauss, H is the (total) field in oersteds, and M is the magnetization in electromagnetic units. In this equation, B, H, and M are all field vector quantities and addition is performed vectorially.

The M field, the H field, and the B field of a uniformly magnetized bar magnet are shown in Fig. 1.8. In all the cases, the scheme adopted for showing, or plotting, the field in the plane of the paper is the same. At all points, the lines and arrows show the direction of the particular field quantity. The spacing between the lines is inversely proportional to the field magnitude. The closer the lines, the higher the field strength. In this scheme, which is identical to that used to depict, for example, airflows, the "flow" or "flux" of the field quantity between all adjacent pairs of lines is equal. In dynamic fields, such as airflows, the lines are called *streamlines*. The B, H, and M fields are static, but the same nomenclature

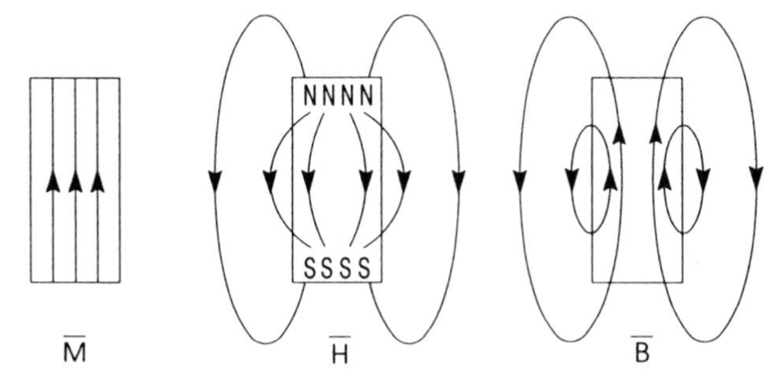

Fig. 1.8. The magnetization, magnetic field, and flux density fields of a bar magnet.

persists. The streamlines are also called *lines of force,* and between them, for the B field, flows the magnetic flux ϕ where

$$\phi = \int B \cdot dA. \tag{1.15}$$

The M field pattern of Fig. 1.8 shows parallel lines within the uniformly magnetized magnet only. The magnetization outside the magnet is zero. The H field drawing shows the magnetic poles caused by the changing magnetization on the ends of the magnet. Where M is decreasing, north, or positive, poles arise as at the top of the magnet. The convention is that H lines of force emanate from the north, or positive, poles. Note that inside the body of the magnet, the magnetic field looks very similar to the electric field between two electrically charged capacitor plates and is not uniform. This is because the bar magnet's shape is not that of an ellipsoid of revolution.

The magnetic field inside the magnet generally opposes the magnetization and is, appropriately enough, called the *demagnetizing field.* The magnetic field outside the magnet, which arises from the very same magnetic poles causing the demagnetizing field inside, is often called the *fringing field.* Because both have the same origin, it is clear that large external fringing fields imply high internal demagnetizing fields.

The B field plot is, of course, identical to that of the H field for all the points outside the magnet. This is because in free space, $B = H$, there being no magnetization. It follows that it is immaterial whether one speaks of the field in the gap of an electromagnet or a magnetic writing head as being B in gauss or H in oersteds.

Inside the magnet, the vector addition of H and $4\pi M$ produces the converging flow shown in Fig. 1.8. Note, particularly, that unlike the H and M fields, the B field flows continuously and has no sources or sinks. This is a consequence of the fact that isolated magnetic monopoles do not exist. North and south poles coexist in equal numbers as dipoles.

Further Reading

Additional reading material is listed here that will prove helpful to readers who seek more detailed information.

Bozorth, Richard M. (1951), *Ferromagnetism,* Van Nostrand-Reinhold, Princeton, New Jersey. (Available from University Microfilms.)

Jiles, David (1991), *Introduction to Magnetism and Magnetic Materials,* Chapman and Hall, London.

Smit, J., and Wijn, H. P. J. (1959), *Ferrites,* Wiley, New York.

CHAPTER

2

The Writing Process

In this chapter, the fundamentals of the writing process are reviewed. This review is of great importance because some aspects of the design of magneto-resistive heads depend on the written magnetization. In order to obtain the maximum output voltage form a magneto-resistive reading head, it *must match properly* the written magnetization transition in the recording medium. The topic of this proper matching is pursued extensively in Chapter 14.

Most applications of MRHs are in digital recording. Accordingly, this chapter deals only with the writing process in digital or binary recording.

Write Head Field

The stylized writing head, shown in Fig. 2.1, consists of three parts: the core, the coil, and the gap.

In ferrite heads the core is usually made of NiZn or MnZn ferrite. These materials are electrical insulators which can be operated at high frequencies (>10 MHz) without requiring thin laminations. In the thin-film heads, the core structure is usually called the yoke. They are made of thin layers of permalloy (81Ni/19Fe) or AlFeSil (an aluminum, iron, and silicon alloy) in, typically, 2- to 4-μm thicknesses.

The coil carries the writing current, I_w, which is typically of magnitude 10 to 20 mA peak. The writing current is toggled from one polarity to the other in order to write digital transitions of the remanent magnetization in the recording medium. Understanding the nature of these transi-

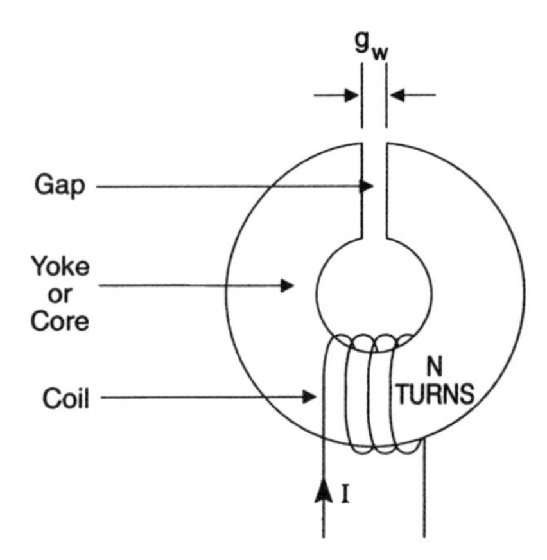

Fig. 2.1. An idealized writing head showing the core or yoke, the coil, and the gap.

tions is the principal purpose of this chapter. The inductance of a head depends on the square of the number of turns in the coil. Because high-voltage write amplifiers are required in order to drive high-inductance heads, the normal design trend is to use a small number of turns and high current rather than vice versa.

The gap, g_w, permits the magnetic flux circulating in the core to fringe out and intercept the recording medium as shown in Fig. 2.2. When the head-to-medium spacing and medium thickness are held constant, changing the write gap length g_w has little effect on the writing process. This is because the coil current I_w is changed simultaneously, so that the same value of the deep gap field, H_0, is maintained. The deep gap field H_0 in oersteds is

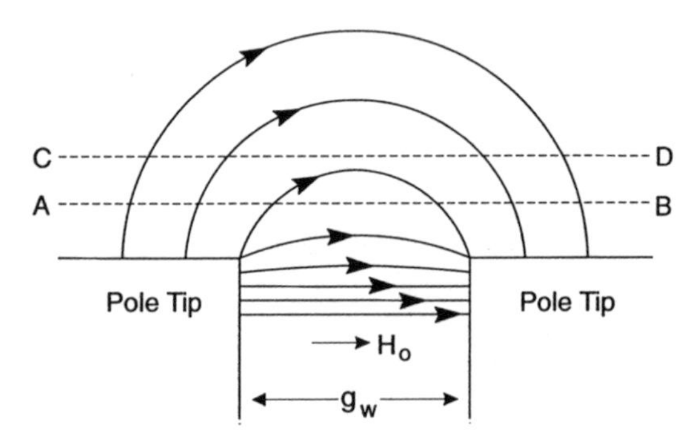

Fig. 2.2. The fringing field around the top of the gap of a write head.

$$H_0 = \frac{0.4\pi NI}{g_w}, \tag{2.1}$$

where N is the number of turns on the coil, I is the writing current in amperes, and g_w is the gap length in centimeters.

In high-density recording, the deep gap field required is

$$H_0 = 3H_c, \tag{2.2}$$

where H_c is the coercivity of the recording medium as shown in Fig. 2.3. To prevent appreciable magnetic saturation in the pole tips, it is necessary that

$$H_0 \leqslant 0.6B_s, \tag{2.3}$$

where B_s is the saturation flux density of the pole or yoke material.

As long as pole tip saturation is avoided, the horizontal component, H_x, of the fringing field above the gap at a point P is

$$H_x = \frac{H_0}{\pi}\tan^{-1}\left(\frac{yg_w}{x^2 + y^2 - g_w^2/4}\right), \tag{2.4}$$

where x and y are the horizontal and vertical coordinates of the point P. The origin of the coordinates is at the top and on the centerline of the gap. In Fig. 2.4 the horizontal component H_x is plotted versus horizontal distance x for two cases where the point P travels across the top of the write head along trajectory A–B and C–D. Note that the trajectory closer to the head (A–B) has both a higher maximum field and higher field gradients, dH_x/dx.

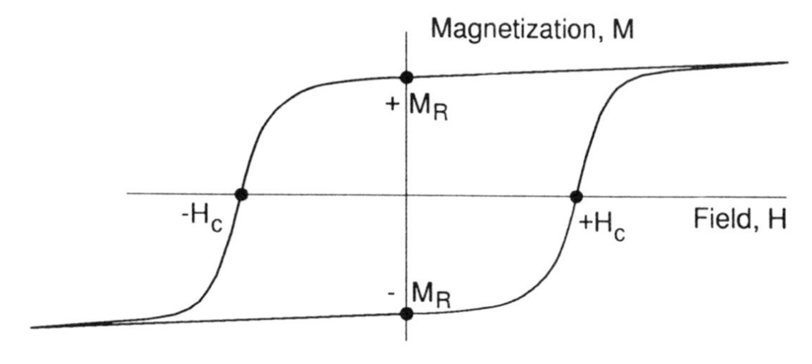

Fig. 2.3. The magnetization versus field hysteresis loop of a recording medium.

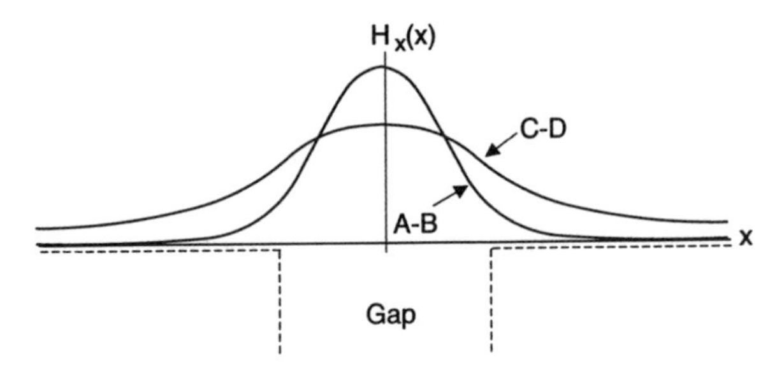

Fig. 2.4. Plots of the horizontal component of the fringing field versus distance along the top of the head.

Written Magnetization Transition $M(x)$

When the write current is held constant, the magnetization written in the recording medium is at one of the remanent levels, M_R, shown in Fig. 2.3. When the write current is suddenly changed from one polarity to the other, the written magnetization undergoes a transition from one polarity of remanent magnetization to the other.

Here it is assumed that the transition has the form shown in Fig. 2.5,

$$M(x) = \frac{2}{\pi} M_R \tan^{-1}\left(\frac{x}{f}\right), \tag{2.5}$$

where f is the so-called "transition slope" parameter. Note that, as shown in Fig. 2.5, the maximum slope, which occurs at $M = 0$, is $2M_R/\pi f$ and

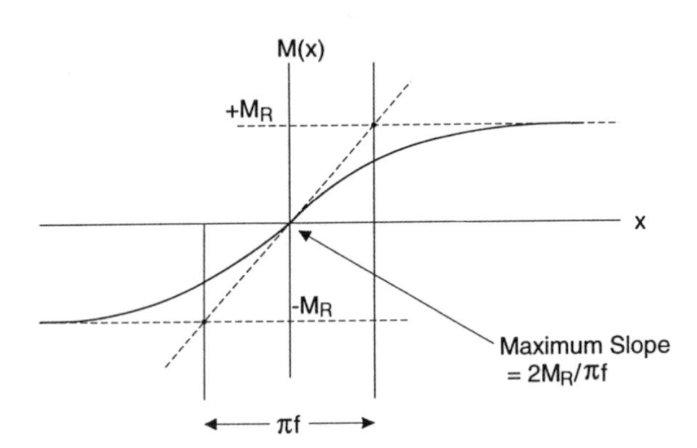

Fig. 2.5. The arctangent magnetization transition showing the maximum slope and the approximate width.

that a straight-line approximation of the transition has a width of πf. As f is reduced, the transition becomes steeper and more closely approaches the binary ideal of the step function with $f = 0$.

The geometry of the write process is shown in Fig. 2.6. The write head-to-medium spacing is d and the recording medium thickness is δ. The gap length between the two poles P1 and P2 need not be specified for the reason already discussed.

The write current is adjusted so that the horizontal component of the fringing field on the midplane of the recording medium meets a specific criterion. This criterion, shown in Fig. 2.7, is that the horizontal position, where the field magnitude is equal to H_c, must coincide with the position where the head field gradient, dH_x/dx, is greatest. Meeting this criterion sets both the magnitude of the write current I_w and the horizontal position of the center of the magnetization transition $[x = (d + \delta/2)/\sqrt{3}]$. The maximum head field gradient is

$$\left(\frac{dH_x}{dx}\right)_{\max} = \frac{\sqrt{3}}{2}\,\frac{H_c}{(d + \delta/2)}\,. \tag{2.6}$$

The written magnetization transition, M versus x, is shown again in Fig. 2.8. Also shown is the magnetic pole density, ρ versus x, and the demagnetizing field on the centerline of the medium, H_d versus x. Note that because the magnetization increases through the transition, the pole density is negative and has south polarity. The maximum slope or gradient of the demagnetizing field is

$$\left(\frac{dH_d}{dx}\right)_{\max} = \frac{-4M_R\delta}{f^2}\,. \tag{2.7}$$

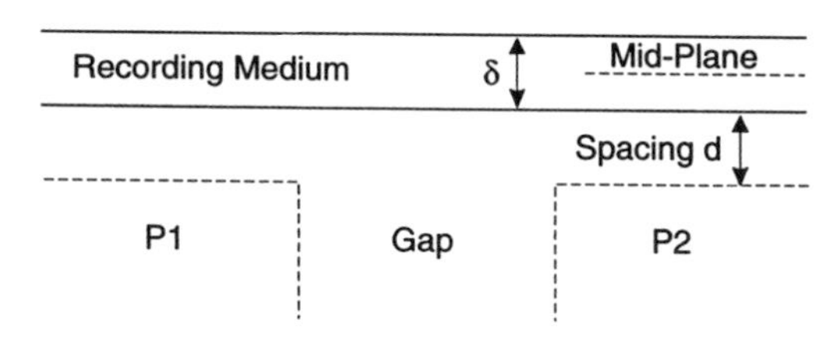

Fig. 2.6. The geometry of the writing process, showing the head-to-medium spacing and the midplane of the thin recording medium.

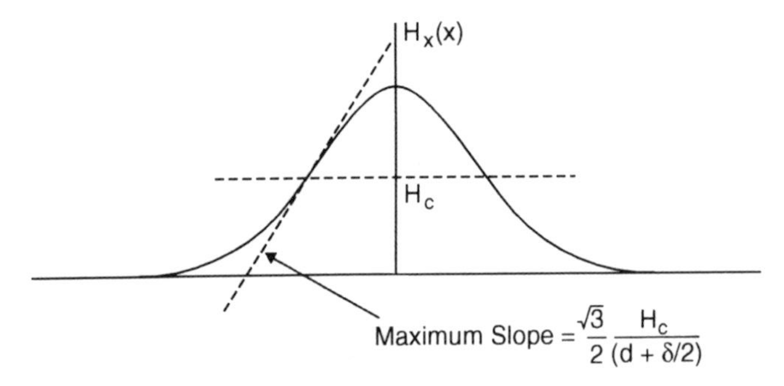

Fig. 2.7. The optimized write head field versus distance showing that the maximum slope occurs at the coercivity.

The maximum gradient occurs at the center ($M = 0$) of the transition. The smaller the value of the slope parameter f, the higher the magnitude of the demagnetizing field and its gradient.

In the Williams and Comstock model of the digital write process, the "slope" equation is used, where

$$\frac{dM}{dx} = \frac{dM}{dH} \cdot \frac{d(H_{\text{head}} + H_{\text{demag}})}{dx}. \tag{2.8}$$

For a square loop recording medium, dM/dH is very high, and a convenient approximation is to set the maximum head field gradient equal to the maximum demagnetizing field gradient. Upon setting Eqs. (2.6) and (2.7) equal, the result is

$$f = 2\left[\frac{2}{\sqrt{3}} \frac{H_c}{M_R} \delta(d + \delta/2)\right]^{\frac{1}{2}}. \tag{2.9}$$

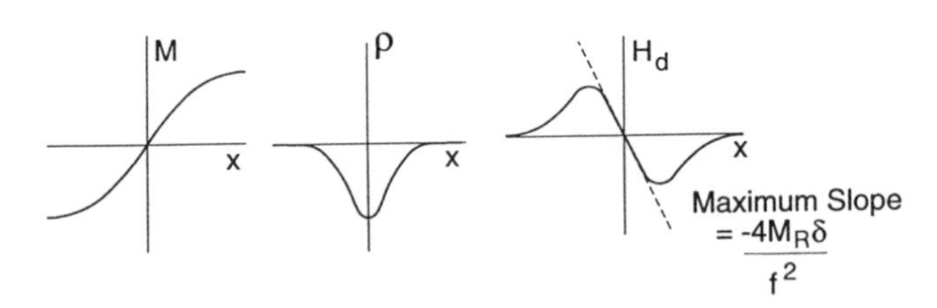

Fig. 2.8. Plots of the magnetization, the pole density, and the demagnetizing field of a written transition.

The writing problem is now completely solved because f is but the single parameter required to define fully the magnetization transition of Eq. (2.5). Note that possible ways to reduce the transition width, by reducing f, are to use higher coercivities, lower remanences, smaller flying heights, and thinner media. With the exception of lowering the remanence, all have been exploited in the past. When inductive reading heads are used, reducing the remanent magnetization is not an acceptable strategy, however, because it always reduces the signal and signal-to-noise ratio of the recorder. With MRHs, however, lower remanence discs may be used as is explained at the end of Chapter 14.

Equation (2.9) for the transition slope parameter f is used in the simplified design of a shielded magneto-resistive head treated in Chapter 14.

Further Reading

Additional reading material is listed here that will prove helpful to readers who seek more detailed information.

Williams, M. L., and Comstock, R. L. (1971), "An Analytical Model of the Write Process in Digital Magnetic Recording," in *17th Annual Conf. Proc.*, pp. 738–742, American Institute of Physics (also in R. M. White; see Bibliography).

CHAPTER

3

The Reading Process

The reading process is reviewed in this chapter for several reasons. First, it is necessary to know the fringing fields surrounding the written magnetization in the recording medium in order to understand the behavior of unshielded magneto-resistive heads (MRHs). Second, when the magnetic flux flowing around a normal inductive reading head is known, then it is possible to deduce how shielded MRHs work in an extremely elegant way. Third, a knowledge of the output voltage spectrum and the isolated magnetization transition pulse shape of an inductive head facilitates direct comparisons with the equivalent properties of MRHs.

Recording Medium Fringing Fields

The fringing field or flux of a recorded medium is shown in Fig. 3.1. The written magnetization waveform is indicated as the dashed line. Where this magnetization decreases, north poles occur and vice versa for south poles.

The magnetic field and flux fringes equally above and below the

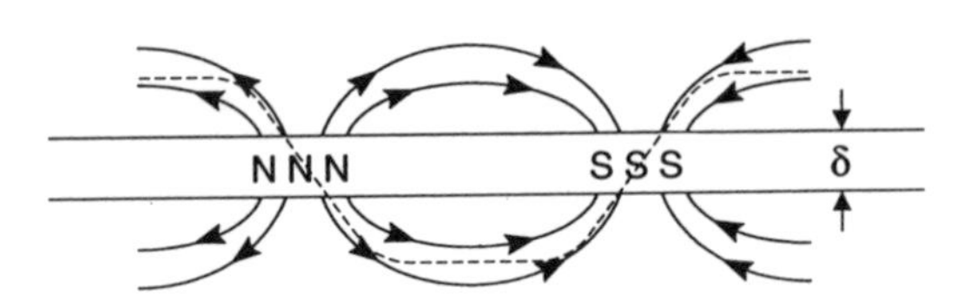

Fig. 3.1. The fringing field above and below a recording medium with a written magnetization pattern (dashed line).

medium, flowing from the north to the south poles. Suppose the written magnetization waveform is

$$M_x(x) = M_R \sin kx, \tag{3.1}$$

where M_R is the maximum amplitude of the written magnetization and k is the wavenumber ($=2\pi/\lambda$), where λ is the sinusoidal wavelength. The horizontal and vertical components of the fringing field at a point (x,y) below the medium are

$$H_x(x,y) = -2\pi M_R(1 - e^{-kd})\, e^{-ky} \sin kx, \tag{3.2A}$$

and

$$H_y(x,y) = 2\pi M_R(1 - e^{-kd})\, e^{-ky} \cos kx. \tag{3.2B}$$

These expressions for H_x and H_y will be used in Chapter 7 for the analysis of Hunt's unshielded MRHs.

Read-Head Flux

When a reading head made with relatively massive pole pieces of highly permeable material is placed close to the recording medium, the field and flux flow patterns are changed profoundly as shown in Fig. 3.2. Relatively massive here means large compared to the recorded wavelength λ. With the reading head in place, almost all the fringing flux flows into the pole pieces, P1 and P2, due to their "keepering" action.

The reading head has the same structure as that of the writing head discussed in the last chapter. Because the pole pieces have a high magnetic permeability ($\mu = dB/dH$), most of the fringing flux flows deep in the head passing through the coil. Very little flux flows through the nonmagnetic gap. The ratio of the flux passing through the coil to the flux entering the top surface of the head is called the *read-head efficiency*. In inductive heads, the efficiency is normally in excess of 80%.

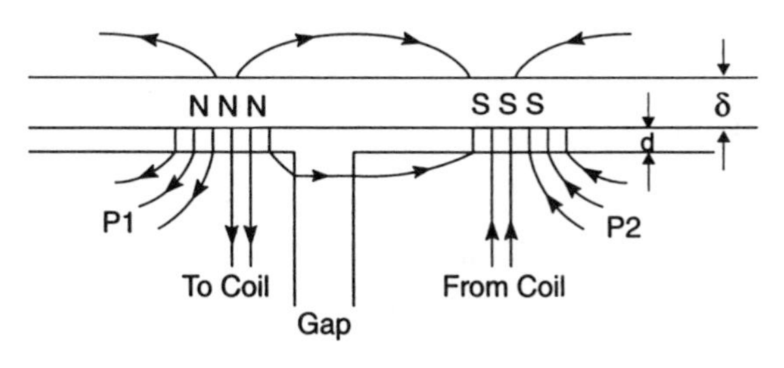

Fig. 3.2. The flux flow around the top of the gap in a highly permeable reading head.

For the sinusoidal recording the flux ϕ threading the coil of a 100% efficient reading head is

$$\phi(x) = -4\pi M_R W \frac{(1 - e^{-k\delta})\, e^{-kd}}{k} \frac{\sin kg/2}{kg/2} \sin kx, \qquad (3.3)$$

where d is the head-to-medium spacing, g is the gap length, and w is the track width.

Output Voltage

The time rate of change of the flux linkages, $N\phi$, in a head coil with N turns is proportional to the read-head's output voltage, E. Thus,

$$E = -10^{-8} \frac{d(N\phi)}{dt} = -10^{-8} N \frac{d\phi}{dt} = -10^{-8} NV \frac{d\phi}{dx} \qquad \text{[volts]}, \quad (3.4)$$

where V is the head-to-medium relative velocity. On putting Eq. (3.3) into Eq. (3.4), the well-known result is

$$E(x) = 10^{-8} NVW4\pi M_R(1 - e^{-k\delta})\, e^{-kd} \frac{\sin kg/2}{kg/2} \cos kx. \qquad (3.5)$$

Note that the output voltage is proportional to the number of coil turns N, the head-to-medium velocity V, and the written remanence M_R.

The term in parentheses in Eq. (3.5) is called the *thickness loss* and it shows that the read head is unable to sense magnetization patterns written deep into the medium. The exponential term e^{-kd} is called the *spacing loss* and it is often quoted as $-55d/\lambda$ dB.

The factor $\sin kg/2/(kg/2)$ is called the *gap loss*. At the first gap null, at wavelength $\lambda = g$, the gap loss term is equal to zero. The fact that the output voltage waveform is a cosine when a sine wave is written shows that the phase of the output signal is lagging 90° behind the written magnetization.

Output Spectrum

The peak magnitude of the output voltage as a function of frequency is called the *spectrum*. The temporal frequency $f = V/\lambda$ and the angular frequency $\omega = 2\pi f$. The spatial frequency or wavenumber $k = 2\pi/\lambda = \omega/V$ so that $\omega = Vk$.

The spatial frequency spectrum corresponding to Eq. (3.5) is just

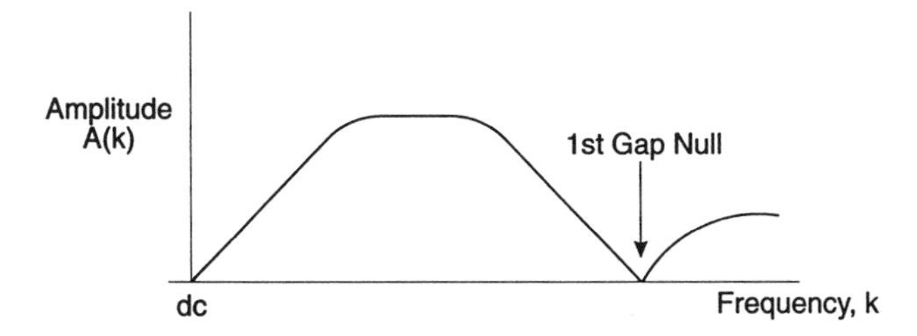

Fig. 3.3. The amplitude of the output voltage spectrum versus wavenumber showing dc and gap nulls.

$A(k) = E(x)/\cos kx$ and it is shown in Fig. 3.3. Note that it has zeros at both dc and the first gap null.

Digital Output Pulse

When the reading head passes over a written magnetization transition of the type discussed in Chapter 2, the coil flux and output voltage are as shown in Fig. 3.4.

The Fourier transform of the magnetization transition

$$M(x) = \frac{2}{\pi} M_R \tan^{-1}\left(\frac{x}{f}\right) \tag{3.6}$$

is

$$M(k) = 2M_R e^{-jkf}/jk. \tag{3.6b}$$

Here f is the transition slope parameter and $j = \sqrt{-1}$. The Fourier transform, $M(k)$, is simply the spatial frequency domain representation of the x domain transition $M(x)$.

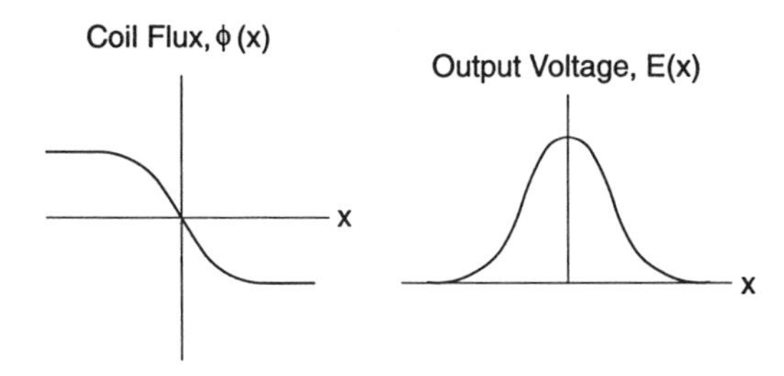

Fig. 3.4. The flux flowing in and output voltage of a head reading a magnetization transition.

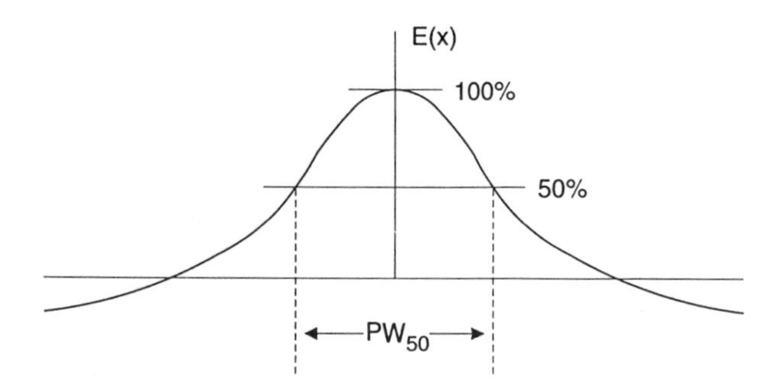

Fig. 3.5. The isolated transition output pulse showing the inevitable negative undershoots.

Because the inductive head reading process is strictly linear, the output signal spectrum when reading the magnetization transition can be written immediately:

$$0(k) = M(k)E(k), \tag{3.7}$$

where $E(k) = jA(k)$ is the read-head output voltage spectrum. The digital output pulse, $0(x)$, is the inverse Fourier transform of Eq. (3.7), and is shown in Fig. 3.5. Note that the pulse is symmetrical for $\pm x$. The negative "undershoots" shown are a direct consequence of the fact that the a reading head, regardless of whether it is inductive or magneto-resistive, cannot have a dc response.

When the read-head gap length g is small, the output pulse is

$$0(x) = 10^{-8}\, NVW4\pi M_R \log_e \frac{(d + f + \delta)^2 + x^2}{(d + f)^2 + x^2}. \tag{3.8}$$

The 50% pulse width, PW_{50}, measured at 50% of the peak amplitude, is

$$PW_{50} = 2[(d + f + \delta)(d + f) + g^2/4]^{1/2}. \tag{3.9}$$

Equations (3.8) and (3.9) are used widely later in this book.

Further Reading

Additional reading material is listed here that will prove helpful to readers who seek more detailed information.

Wallace, R. L. (1951), "The Reproduction of Magnetically Recorded Signals," *Bell Syst. Tech. J.* (also in R. M. White; see Bibliography).

CHAPTER

4

The Anisotropic Magneto-Resistive Effect

The Basic Effect

In 1857, William Thomson, who later became Lord Kelvin, discovered that the electrical resistance of an iron bar changed as its magnetization was altered. This phenomenon is now called the *anisotropic magneto-resistive* (AMR) effect and it remained a mere academic curiosity until Bob Hunt invented the magneto-resistive head (MRH).

In a thin film of magnetic material, the magnetization can be single domain and not be subdivided by domain walls into the multidomain state. Moreover, in a thin film, demagnetizing fields force the magnetization direction to remain virtually parallel to the plane of the film.

In the thin-film single-domain state, the AMR effect can be described very simply. With reference to Fig. 4.1, when the angle between the magnetization, M, and the electrical measuring or sensing current, I, is θ, the electrical resistance, R, can be written

$$R = R_0 + \Delta R \cos^2 \theta, \tag{4.1}$$

where R_0 is the fixed part and ΔR is the maximum value of the variable part of the resistance, respectively.

In Fig. 4.1, note that the thickness of the thin film is T, the film depth is D, and the film width is W. This labeling of the MR sensor or MR

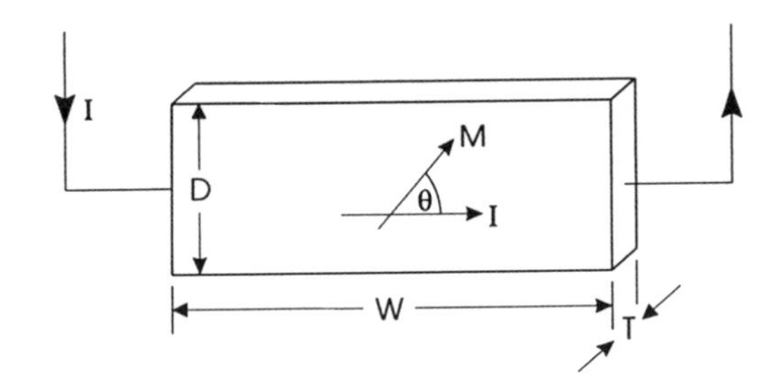

Fig. 4.1. The AMR effect in a single-domain thin-film strip.

element (MRE) is used consistently throughout this book. The width direction is usually both the electrical current direction and the recorded track-width direction.

A plot of the variable part of the resistance of a material like permalloy versus the magnetization angle is shown in Fig. 4.2. Because the maximum value of the variable part of the resistance occurs at $\theta = 0$, it is seen that, perhaps counter to intuition, the resistance is maximum when the magnetization and current are parallel. Note that the resistance change versus angle characteristic has an inflection point at $\theta = 45°$ where the characteristic is, for small angular excursions, almost linear.

Magneto-Resistive Sensors or Elements

The fundamental idea in all MRHs is that the magnetic field produced by the written or recorded magnetization pattern in the tape or disk rotates the MRE magnetization angle. When the magnetization angle in the MRE is properly biased with a vertical bias field, produced by the means discussed in Chapter 5, small changes in resistance, δR, are almost directly proportional to small amplitude fields from the recording medium.

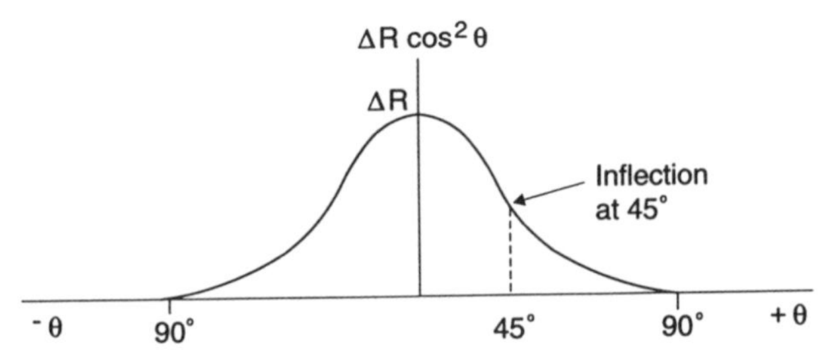

Fig. 4.2. The magneto-resistive change in resistance versus magnetization angle.

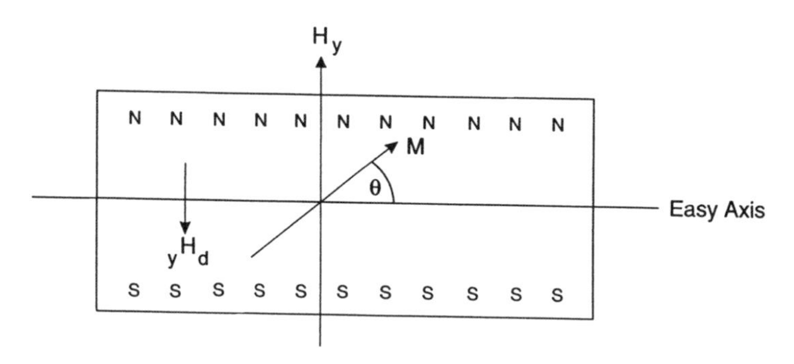

Fig. 4.3. An MRE showing the easy axis, the disk field, the magnetic poles at the edges, and the vertical demagnetizing field.

In all magneto-resistive heads, the output signal voltage is, by Ohm's law, equal to the resistance change times the measuring current: $\delta V = I\delta R$.

When the magnetization angle θ in the MRE increases, magnetic poles are generated in the top and bottom regions of the sensor, as shown in Fig. 4.3. These poles, of north polarity at the top and south at the bottom, generate an internal field, usually called the *demagnetizing field*, ${}_yH_d$. If the vertical magnetic field, H_y, which is the sum of the vertical bias field and the tape or disk field, is acting up, the demagnetizing field induced by the magnetization angle θ is down as shown in Fig. 4.3. Whereas the vertical field acts to increase θ, the demagnetizing field acts to reduce θ. The value of the demagnetizing field, averaged over the element depth D, is proportional to T/D. Thus, the thicker and shallower the MRE, the greater the demagnetizing fields for a given magnetization angle θ.

Figure 4.4 shows the change in resistance versus vertical field, H_y. If the demagnetizing fields are negligible (thin and deep element), the magnetization angle is

$$\theta = \sin^{-1}(H_y/H_k) \tag{4.2}$$

and the resistance is

$$R = R_0 + \Delta R[1 - (H_y/H_k)^2]. \tag{4.3}$$

Here H_k is the fictitious "anisotropy field" discussed later in this chapter. The variable part of the resistance is parabolic, shown as a dashed line in

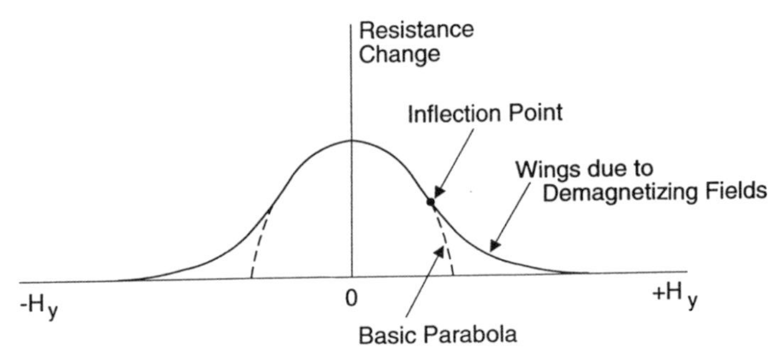

Fig. 4.4. The magneto-resistive change in resistance versus disk field, showing the effects of the demagnetizing field.

Fig. 4.4, and has no inflection point. The effect of taking demagnetizing fields into account is the appearance of the wings shown as solid lines. More vertical field is required to achieve a given magnetization angle. Note that an inflection point once more exists. At the inflection point, the magnetization angle might typically vary almost 0° at the top and bottom of the element to perhaps 60° at the middepth.

Provided the vertical field is not sufficient to saturate the MRE, that is, $M_y < M_{\text{sat}}$ at the middepth $y = D/2$, an exact analytical solution for the magnetization angle θ is known for the usual case where the anisotropy field H_k is negligible compared with the demagnetizing field ${}_yH_d$. The magnetization angle as a function of element depth y is

$$\theta(y) = \tan^{-1}\left\{\frac{H_{\text{bias}}}{2\pi M_s T}\left[\left(\frac{D}{2}\right)^2 - y^2\right]^{\frac{1}{2}}\right\}. \tag{4.4}$$

For many purposes, it is both convenient and sufficient to approximate the change in resistance versus vertical field characteristic with the straight heavy solid line shown in Fig. 4.5. The slope of this approximated characteristic is equal to $-\frac{1}{2}\Delta R/{}_yH_{\text{bias}}$ and it represents the sensitivity of the MRE when vertical bias is used. This approximation was made in the first analysis of Hunt's unshielded MRHs, the results of which are discussed in Chapter 7.

When no vertical bias is used, the sensitivity, or slope, of the resistance change versus field characteristics is, of course, zero.

Figure 4.6 depicts the small resistance changes δR in response to oscillating polarity small fields from a written magnetization pattern in a tape or disk. When the proper vertical bias field is used, the output

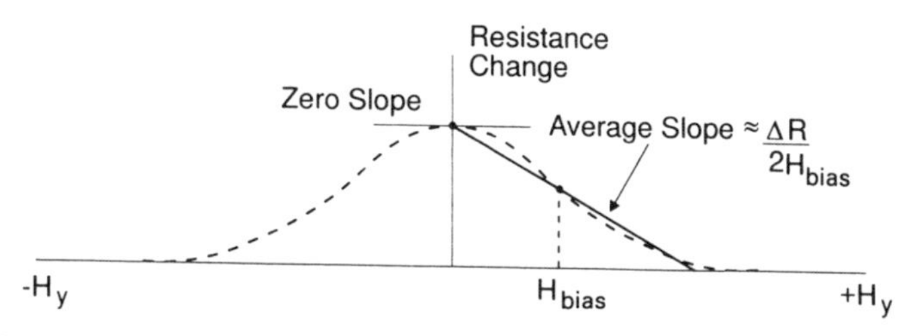

Fig. 4.5. The MR resistance change versus the disk field characteristic, showing the average slope and the bias field.

voltage, $I\delta R$, is both large and almost linear. Typically, deviations from linearity cause about -20 dB of even harmonic distortion, which, though satisfactory for a binary or digital data channel, is not sufficiently linear for an analog signal channel. On the other hand, if vertical biasing is not used, the response is of low sensitivity and is highly nonlinear, behaving as a full-wave rectifier.

It is important to realize that even when the optimum vertical bias is used, the output of an MRE is inherently nonlinear even for small tape or disk fields. In this respect an MRH differs greatly from the usual inductive reading head. Accordingly, simple MRHs always show pulse amplitude asymmetry and even harmonic distortion so that the positive and negative output pulses are of unequal amplitude. Furthermore, the PW_{50}s of the opposite polarity output pulses must be unequal. Several schemes for removing this inherent asymmetry are discussed in Chapters 5 and 10 of this book.

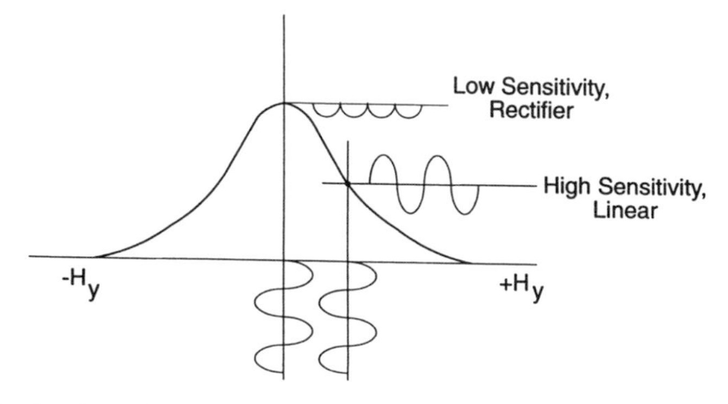

Fig. 4.6. A schematic representation of the MR reading of disk fields with and without a bias field.

The precise origins of both the odd and even harmonic distortion can be understood with the help of Fig. 4.7. Suppose the upper portion of the resistance change characteristic labeled (a) is folded or reflected abaout an imaginary horizontal plane to form the dashed line marked (b). Let the reflected part (b) be folded again, about an imaginary vertical plane, to form the second dashed line (c).

Now if the twice-folded dashed line (c) happened to coincide with the lower portion of the characteristic labeled (d), odd harmonic distortion only would occur in the MRH response. Odd harmonic distortion simply causes the output pulses to be clipped or limited and does not cause pulse amplitude asymmetry.

In reality, however, dashed section (c) does not match the lower part of the characteristic (d). The mismatch causes the generation of even harmonics and the output pulses to be of unequal amplitudes.

The mismatch shown in Fig. 4.7 causes a negative H_y pulse to have a larger amplitude than that of a positive H_y. This is because part (a) has a greater negative slope than does part (d).

All present MRHs use either vertical bias or other means to achieve quasi-linear operation. Most of the schemes which have been proposed to achieve vertical biasing are discussed in Chapter 5.

Just as the finite depth D of the MRE causes the vertical demagnetizing fields ${}_yH_s$, as already discussed, the finite width W causes a horizontal demagnetizing field, ${}_xH_d$. This is shown in Fig. 4.8.

Both the vertical and horizontal demagnetizing fields depend on the MRE dimensions and magnetization angle θ, and they vary spatially. Plots showing this variation and the average value of the vertical and

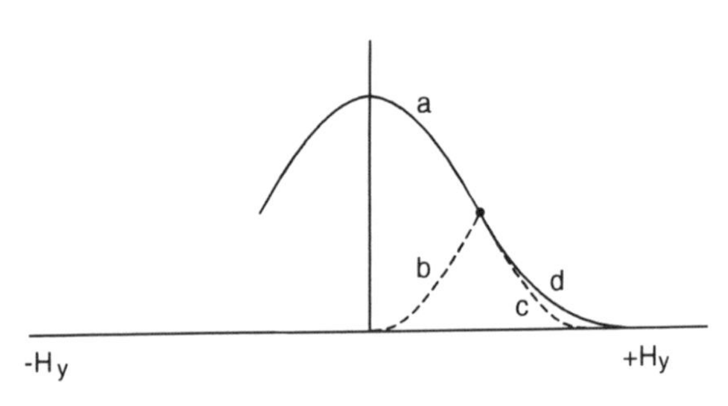

Fig. 4.7. The MR resistance change versus field characteristic, showing the lack of symmetry about the bias point.

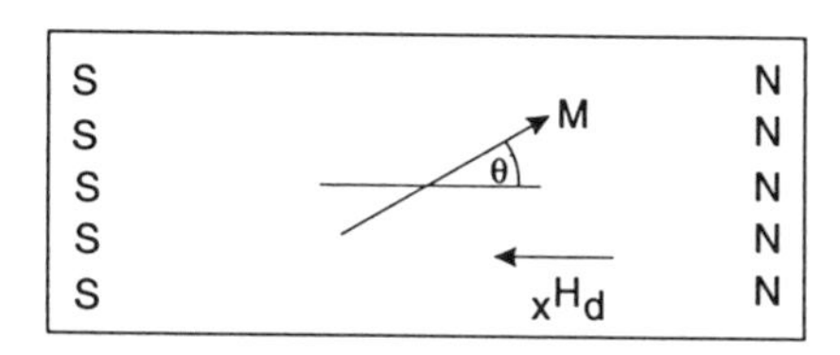

Fig. 4.8. A MR element showing the end-zone poles and the horizontal demagnetizing field.

horizontal demagnetizing fields are shown schematically in Figs. 4.9 and 4.10, respectively.

To keep the single-domain magnetization in the MRE stable and free from hysteresis, it is usually necessary to provide a horizontal bias field in addition to the vertical bias field already discussed. The existence of hysteresis causes the resistance change for increasing magnitude vertical fields to be different from that for decreasing vertical fields. The presence of hysteresis indicates that, in some part of the MRE, usually at the ends, there are domain walls which have moved discontinuously from one pinning site to another. This discontinuous motion causes jumps in the MRE magnetization angle, resistance, and output voltage. These jumps are usually termed Barkhausen noise. It is essential that hysteresis and Barkhausen jumps be eliminated, and several of the means proposed for achieving this are discussed in Chapter 6.

Magneto-Resistive Coefficient $\Delta\rho/\rho_0$

The electrical resistance, R, of the MRE, in terms of the resistivity, ρ, of the material, is

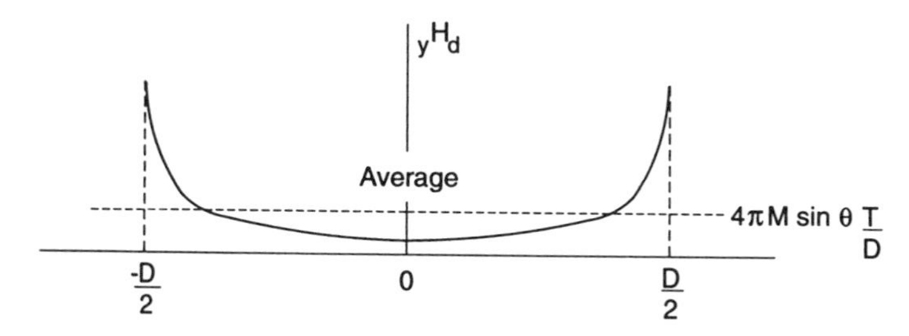

Fig. 4.9. A plot of the vertical demagnetizing field versus depth, showing the average value.

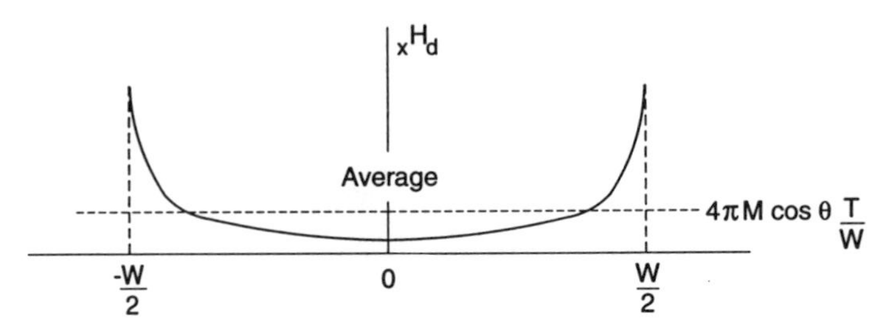

Fig. 4.10. A plot of the horizontal demagnetizing field versus width, showing the average value.

$$R = \rho W/TD. \tag{4.5}$$

In magneto-resistive materials

$$\rho = \rho_0 + \Delta\rho \cos^2 \theta, \tag{4.6}$$

where ρ_0 is the fixed part and $\Delta\rho$ the maximum value of the variable part of the resistivity.

The magneto-resistive coefficient is defined to be $\Delta\rho/\rho_0$ and it is the maximum fractional change in the resistivity. For many purposes, this coefficient serves as a figure of merit of the MR material. Note that $\Delta\rho/\rho_0$ equals $\Delta R/R_0$.

Permalloy is a binary alloy of nickel and iron of nominal composition 81Ni/19Fe. The magneto-resistive coefficient of permalloy is approximately 4% in thick (>1000-Å) films. In thinner films, the coefficient decreases due to the increasing importance of surface scattering of the conduction electrons, which mainly increases the fixed part of the resistivity. For the 2- to 300-Å-thick films frequently used today, the magneto-resistive coefficient is only about 2%.

It is known that the AMR originates in the anisotropic scattering of *d* shell conduction electrons in the exchange split *d* bands. As was discussed in Chapter 1, the *d* shells contain the electrons which are responsible for ferromagnetism and ferrimagnetism. Exchange split merely means that the energy of the up and down spin conduction electrons differs by the quantum-mechanical exchange energy discussed in Chapter 1.

Seemingly convincing analyses of the AMR are frequently given in terms of the electron density of states diagram and the Fermi level. This is, of course, conventional in explanations of the semiconductor behavior of, say, gemanium and silicon. The Fermi level is the most probable maximum energy of the conduction electrons.

Unfortunately, however, similar analyses have little practical utility in MR materials other than for elementary pedagogical purposes. The difficulty is that the anisotropic part of the resistance depends on the exact 3-D shape of the Fermi surface and this is not known precisely excepting for a very few magnetic materials. The Fermi surface is the 3-D envelope of the Fermi level. In some materials, the MR coefficient is negative, meaning that the low resistivity state occurs when M and I are parallel. Neither this fact, nor the fact that most materials have positive MR coefficients, can be easily predicted.

In the absence of a viable theory, investigators must resort to experimental work in order to find improved materials. An interesting example of such a search covered no less than 30 binary and 10 ternary alloys (McGuire and Potter, 1975). In this search, it was found that many materials, for example 80Ni/20Co, have extremely high MR coefficients (20%) at liquid helium temperatures. At room temperature, however, only the following four alloys had MR coefficients higher than that of permalloy: 90Ni/10Co (4.9%), 80Ni/20Co (6.5%), 70Ni/30Co (6.6%), and 92Ni/8Fe (5.0%). None of these compositions is, however, suitable for MR sensors due either to high anisotropy or high magneto-striction. The former leads to high coercive force and low permeability, the latter to high sensitivity to mechanical stress and strain.

The Unique Properties of Permalloy

It is indeed remarkable that not only were Hunt's first MRHs made of permalloy but also that permalloy still remains, even 25 years later, the material of choice in both conventional anisotropic MR heads and the proposed giant MR heads discussed in Chapter 13. Moreover, permalloy is also the magnetic material of choice for most thin-film heads.

Before discussing the unique properties of permalloy that give it this dominant role in magnetic recording technology, some other important properties are listed below:

Composition (nominally)	81% Ni, 19% Fe
Saturation magnetization $4\pi M_s$	12,500 G
Electrical resistivity	20.10^{-6} Ω-cm
Magneto-resistive coefficient	2–4%
Thermal coefficient of resistivity	0.3%/°C

The first property of permalloy which is responsible for its dominance in head technology is well known. At compositions close to 81Ni/19Fe, its cubic magneto-crystalline anisotropy, K, passes through zero. Positive K means the easy axes of magnetization are the cube edges, negative K means the easy axes are the cube body diagonals, and zero K means the material is isotropic, having no preferred easy (or hard) axes.

Anisotropy is, of course, due to the spin-orbit coupling of the electron spins to the electronic structure of the material. The greater the change in spin-orbit coupling energy with changes in magnetization direction, the greater the anisotropy. It follows that in noncubic crystal structures such as hexagonal cobalt, the anisotropy is usually much larger than in cubic systems such as iron and nickel.

The fact that $K = 0$ means that in permalloy laminations and thin films produced in the absence of external "orientation" magnetic fields, the coercive force can be very low ($H_c < 1$ Oe) and the permeability very high ($\mu > 1000$). When the material is deposited or annealed in the presence of an "orientation" field, some spatial ordering of the Ni and Fe atoms occurs and the material develops what is called an *induced* uniaxial anisotropy which has its single easy axis parallel to the orientation field. The orientation field has to be sufficient to saturate the magnetization only; higher fields have no further appreciable effect. The induced uniaxial anisotropy is K_u ergs/cm^3 and is, as usual, simply the magnetic energy required, per unit volume, to rotate the magnetization from the easy axis to the hard directions orthogonal to the easy axis. In single-domain thin films, the fictitious anisotropy field, H_k, is equal to $2K_u/M$. The adjective *fictitious* is used because no actual magnetic field exists. However, the actual field required to rotate the magnetization into a hard direction is equal to H_k. In permalloy, $H_k \approx 5$–10 Oe only. The ease with which permalloy can be induced to develop an easy axis during deposition or subsequently by magnetic annealing (typically at 250 to 300°C) is a very useful property in the fabrication of both thin-film inductive heads and magneto-resistive heads.

The second unique property of permalloy is that, at a composition very close to 81Ni/19Fe, the magneto-strictive coefficient λ also passes through zero. Magneto-striction is the phenomenon in which a magnetic material changes its size depending on its state of magnetization. Figure 4.11 shows the effect. If a bar of magnetic material has length l when it is demagnetized ($M = 0$) and has length ($l + \Delta l$) when its magnetization is saturated (M_{sat}), the linear magneto-strictive coefficient, λ_l, is positive

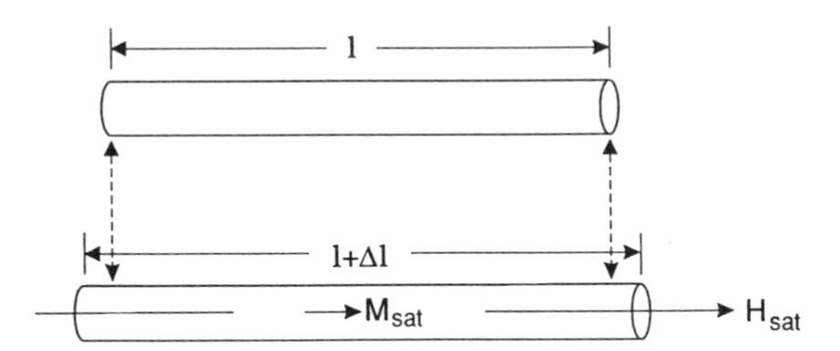

Fig. 4.11. The basic magneto-striction phenomenon showing the length of a bar changing with its magnetic state.

and of magnitude $\Delta l/l$. If the bar shrinks as it becomes magnetized, then λ_l is negative.

The inverse magneto-strictive effect is that the magnetic state of a magnetic material can be changed by the application of external mechanical stress. Thus a magnetized specimen with positive magneto-striction is partially demagnetized by the application of compressive stress. With negative λ_l, tensile stress is required to reduce the magnetization.

Magneto-striction is, of course, another manifestation of spin-orbit coupling. The greater the change in spin-orbit coupling energy with changes in atom-to-atom spacing, the greater the magnetostriction. It is, therefore, likely to be large when the magneto-crystalline anisotropy is large. Indeed, most cobalt alloys have large magneto-striction.

The critical importance of zero magneto-striction in head fabrication arises because it is virtually impossible to produce thin films which are completely stress-free. This is often due to factors related to compositional nonuniformities and gradients. Further, it is almost certain that differential thermal expansion effects between the several materials in the head will cause stresses. Finally, even if it were possible to deposit a stress-free head structure, it is very likely that subsequent mechanical processing of the head (lapping, polishing, etc.) would induce mechanical stresses.

The differential thermal expansion effects caused by I^2R heating in the copper coil of an inductive thin-film write head can give rise to magnetic disturbances when that same head cools down and contracts during subsequent reading operations. The magnetic disturbances associated with the contraction have colorful popular names such as "popcorn" or "snap-crackle" noise and they can be minimized by careful control of the magneto-striction of the magnetic material. Because the coil heating causes tensile stresses in the magnetic yoke structure, it is generally found that making permalloy a little Ni rich, with λ slightly negative, is beneficial for maintaining the proper domain structure in the head.

The fact that, in permalloy, both $K = 0$ and $\lambda = 0$ occur at almost the same composition is responsible for permalloy's dominance in magnetic heads, both thin film and MR, in the past and in the foreseeable future.

When we want the $K = \lambda = 0$ condition to occur simultaneously in permalloy, it is necessary to add a third material to the alloy. In the well-known magnetic shielding material supermalloy, this is achieved by adding 4% molybdenum.

Permalloy has one notable disadvantage in that the material is mechanically soft and is thus not well suited to abrasive head-to-medium contact applications. In such applications, another $K = \lambda = 0$ material, AlFeSil, often called Sendust, is frequently used. AlFeSil is a hard, brittle alloy which resists wear well, but is considerably more demanding to produce. AlFeSil films can only be produced by sputtering, whereas permalloy films can be made by electroplating, evaporation and sputtering.

Further Reading

Additional reading material is listed here that will prove helpful to readers who seek more detailed information.

Hunt, Robert P. (1970), "A Magneto-resistive Readout Transducer" (digest only), *IEEE Trans.* **MAG-6,** 3.

Hunt, Robert P. (1971), "A Magneto-resistive Readout Transducer," *IEEE Trans.* **MAG-7,** 1.

McGuire, T. R., and Potter, R. I., "Anisotropic Magnetoresistance in Ferromagnetic 3D alloys," *IEEE Trans.* **MAG-11,** 4.

Thomson, W. (1858), *Philosophic Magazine,* London.

CHAPTER

5

Vertical Biasing Techniques

In this chapter most of the techniques proposed for applying the required vertical bias field on the magneto-resistive element (MRE) are discussed. As will be evident by noting the very number of techniques that are treated here, this has been an extraordinarily fertile field for invention.

Vertical bias is sometimes called perpendicular bias and, rather confusingly, transverse bias. Horizontal bias, also known as longitudinal bias, is discussed in Chapter 6.

The description of each biasing technique is followed by a short review of its main advantages and disadvantages.

Permanent Magnet Bias

The earliest, and one of the simplest, ideas for providing vertical bias is shown in Fig. 5.1. Here a thin film of permanent magnet (PM) material is deposited adjacent to the MR element. When the permanent magnet is magnetized up, it produces a down bias field in the MRE. The coupling is purely magneto-static. To prevent exchange coupling and electrical shunting of the measuring current by the PM film, a nonmagnetic, electrically insulating film is interposed as shown. The insulator film is typically 100 to 200 Å of SiO_2 or Al_2O_3.

The main advantage of this scheme is that the many years of experience in making thin-film recording media can be applied to the PM film. In the past, PM films have been made of Co–P or Co–Ni–P, whereas

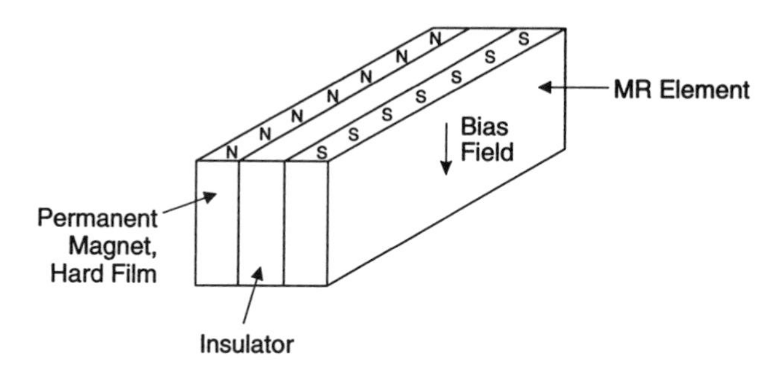

Fig. 5.1. The permanent magnet bias structure.

Co–Cr–Pt is common today. The design problem is also relatively simple. The PM film must have a remanence times thickness product that is approximately equal to $M_sT \sin\theta$ of the MRE.

Disadvantages include the fact that a high magnetic field, of magnitude about equal to the coercivity of the PM material, is present at the top edge immediately adjacent to the tape or disk. It follows that the PM film coercivity must be limited to only 50% of that of the recording medium. Another disadvantage is that the errors in fabrication, which result in an incorrect vertical bias field, cannot be adjusted easily after the MRH has been made. Postfabrication adjustment of the vertical biasing is, on the other hand, easily accomplished in the current biasing schemes discussed next.

Shunt Current Bias

In this early scheme the current flowing into the MRH is divided into two parts. One flows through the MRE as usual. The second part flows through a layer of electrically conductive, nonmagnetic material which is electrically insulated from the MRE. The arrangement is shown in Fig. 5.2.

The current flowing in the MRE cannot provide a vertical bias field in the MRE itself. However, the shunt current in the adjacent layer does cause a vertical bias field in the MRE. By the right-hand rule discussed in Chapter 1, shunt current flowing in the direction shown in Fig. 5.2 produces a down vertical bias field in the MRE.

The magnitude of the shunt current is determined by the voltage applied to the MRH and the resistance of the shunt current conductor film. If the shunt current is I_s and the conductor is of equal depth to the MRE, then the vertical bias field is, over the middepths of the MRE,

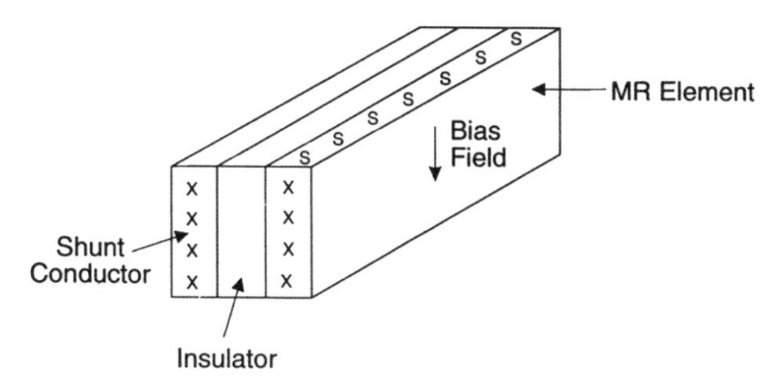

Fig. 5.2. The shunt current bias arrangement.

approximately equal to $0.2\pi I_s/D$ Oe. An estimate can be made of the MRE bias angle using Eq. (4.4).

An advantage of shunt bias is that it requires only minimal materials technology to implement. Moreover, by adjusting the voltage applied across the MRH, the shunt current and thus the vertical bias field can be adjusted after head fabrication.

The main disadvantage of shunt bias is that the effective resistance change is reduced by the shunting when the MRH is operated on a two-terminal device. The effective resistance change is $(R_2/R_1 + R_2)\delta R$ only, where R_1 and R_2 are the MRE and shunt conductor resistances, respectively. Of course, if the MRH is operated as a four-terminal device, the full MRE resistance change δR can be detected at the expense of increased circuit complexity.

A second disadvantage is that vertical bias field magnitude falls off to the approximately $0.1\pi I_s/D$ at the top and bottom ends of the MRE. The critical region of the MRE which is closest to the recording medium is likely to be underbiased. The depth of this underbiased region can be reduced by making the insulator film thinner or by omitting it altogether. When no insulator film is used, however, problems can arise due to diffusion of the conductor layer metal into the permalloy MRE. For example, copper is soluble in permalloy.

Soft Adjacent Layer Bias

The soft adjacent layer (SAL) vertical bias arrangement is shown in Fig. 5.3. Here a film, made of a low-coercivity, high-permeability magnetic material, is placed adjacent to the MRE. To prevent electrical shunting, a thin layer of insulator is interposed.

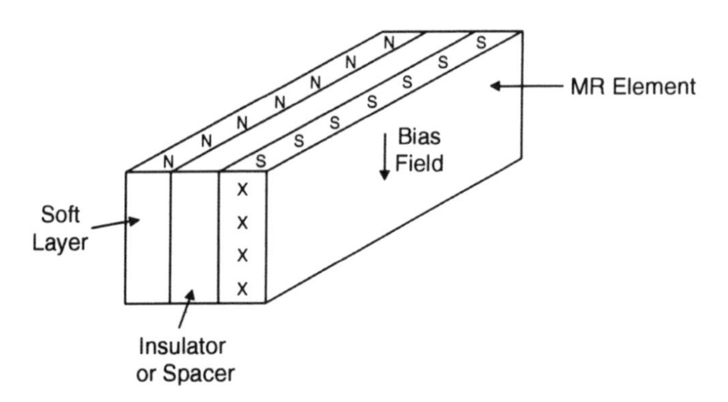

Fig. 5.3. The SAL bias construction.

Typically, the soft film is permalloy, but some designs use other materials, such as AlFeSil or NiFeCr. Insulator layers of SiO_2 and Al_2O_3 are used, with a high resistivity phase of Ta being common in many hard-disk file MRH designs.

The principle of operation of the SAL vertical bias technique is as follows. The current flowing in the MRE produces, by the right-hand rule, a vertical magnetic field in the SAL. This field magnetizes the SAL in the vertical direction. In Fig. 5.3, the MRE current direction shown causes the SAL to be magnetized up. The changing SAL magnetization produces $\rho = -\nabla \cdot M$ magnetic poles, as was discussed in Chapter 1, and these poles generate the desired down vertical bias field in the MRE. Because there is mutual magneto-static interaction between the SAL and the MRE, a simple and accurate way of calculating the vertical bias field in SAL MRHs is not known to this writer. Accordingly the design problem must be solved by using micromagnetic numerical techniques.

The design is usually arranged so that the middepth region of the SAL becomes magnetically saturated. This is done because saturation of the SAL reduces its permeability, which, in turn, reduces the undesirable record medium fringing flux shunting that is inherent in all SAL MRHs. This signal flux shunting occurs because there are two parallel paths for the record medium fringing flux, the SAL and the MRE.

The advantages of SAL MRHs include the fact that adjustments to the vertical bias field can be made by adjusting the current. In principle, the materials technology requirements are minimal.

The SAL designs have two main disadvantages. The first, the signal flux shunting problem, has already been discussed. The second is that correct operation of the head depends on the integrity of the electrical insulating film. If pinholes occur in fabrication or if the SAL and MRE become wiped or smeared together by tape or disk abrasion, the performance is compromised by electrical shunting.

The SAL technique is used by IBM in both tape and disk file MRHs.

Double Magneto-Resistive Element

The double-element MRE vertical biasing scheme is a natural evolution of the shunt bias and SAL bias techniques already discussed.

As shown in Fig. 5.4, the design has two MREs separated by a thin insulator layer. Both MREs carry current, which usually, but not necessarily, flows in the same direction. Current in MRE #1 produces an up vertical bias field in MRE #2. Current in MRE #2 produces a down field in MRE #1.

Because the two MREs interact mutually, a simple way to estimate accurately the magnetization bias angles does not exist.

Advantages include the minimal materials science required, the simple fabrication, and the adjustability of the vertical bias by changing the current magnitudes. Moreover, the media flux shunting problem of the SAL design is avoided because a changing magneto-resistive output signal is available from both MREs.

The two MRE outputs can be sensed either in addition, as in the Kodak Dual-Magneto-Resistor (DMR) head, or in subtraction, as in the Hewlett-Packard Dual-Stripe head. The relative advantages and disadvantages of these arrangements are discussed in Chapter 10.

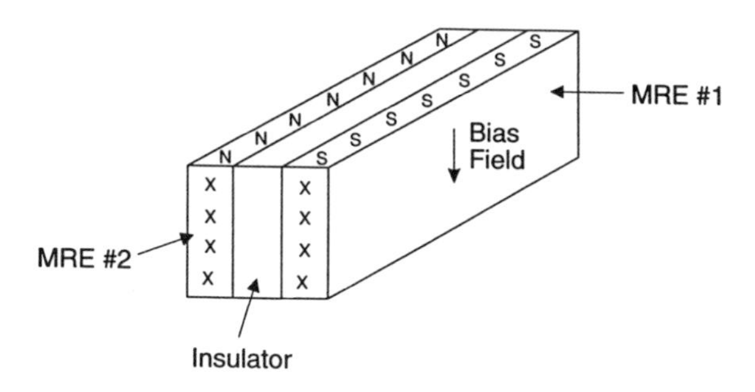

Fig. 5.4. The double-element magneto-resistive bias technique.

Self-Bias

This particularly ingenious vertical biasing technique is applicable to MRHs in which the MRE is placed in the gap between two highly permeable poles or shields. Shielded MRHs are covered in detail in Chapter 8.

Figure 5.5 shows the MRE placed off-center between Shields #1 and #2. The off-center positioning is the essence of the invention.

When current flows in the MRE, it, of course, cannot produce a vertical bias field on itself directly. To meet the magnetic boundary conditions imposed by the high permeability shields, however, a vertical bias field does appear when the MRE is not on the centerline of the gap.

Referring to Fig. 5.5, the effect of the shields' presence is to alter the field produced by the MRE current in a way that looks, in the gap, as though magnetic "images" exist in the shields. In first approximation, these images look like the MRE itself, just as one's own image in a mirror looks like one's self. The images appear to exist at a distance in the shields equal to the MRE-to-shield spacing, just as your own reflection in a mirror appears to be as far behind the glass as you are in front.

Because distance from the MRE to the left-hand image (#1) is smaller than the right-hand image (#2) distance, the vertical bias fields (H_1 and H_2) that each image produces at the MRE are not equal. The net vertical bias field in the case shown in Fig. 5.5 is of magnitude $H_1 - H_2$ and is down.

The main advantage, of course, of self-bias is that it occurs automatically, merely by putting the MRE off-center. No extra structure is required.

The main disadvantage appears to be that the magnitude of the verti-

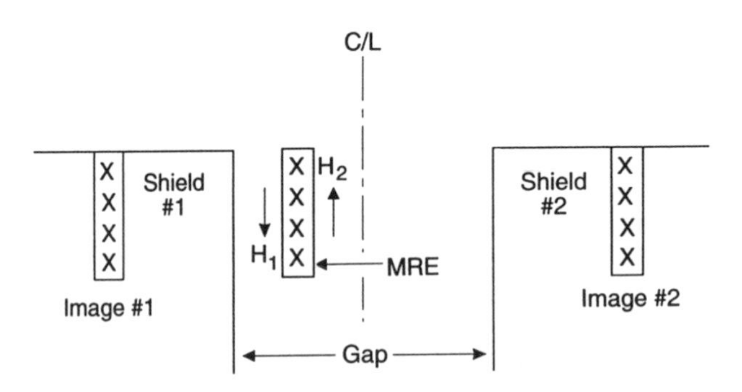

Fig. 5.5. The self-bias method.

cal bias field is often not sufficient. This insufficiency arises because the images are too far away from the MRE and the net bias field is the difference, $H_1 - H_2$. Very often the self-bias vertical bias is augmented by using one of the other techniques discussed in this chapter.

Exchange Bias

Exchange bias, shown in Fig. 5.6, depends on atomic contact between the MRE and another magnetic material. Atomic contact means that every effort is made to ensure that no foreign material or atoms exist in the interface. When the proper interface is achieved, the quantum-mechanical exchange phenomenon, discussed in Chapter 1, couples the adjacent electron spins in the MRE and the other magnetic layer.

The interfacial exchange coupling energy depends on the sub-microscopic details of the interface. The strength of the coupling can be deduced by analysis of the *M–H* loop characteristics of the coupled films. It is usually expressed as an effective exchange field, H_{EX}. When the interfacial exchange coupling energy is J_{EX} ergs·cm^{-2} and the MRE magnetization and thickness are M_R and T respectively, the effective exchange field is $J_{EX}/M_R T$.

The interfacial exchange field can have a profound effect upon the *M–H* loop characteristics of the coupled films. Suppose that the antiferromagnetic layer has magnetocrystalline anisotropy, K ergs·cm^{-3}, and thickness, t. When $J_{EX} < Kt$, the *M–H* loop is shifted unidirectionally because the interfacial coupling is too weak to reverse the spins throughout the entire thickness of the antiferromagnet and this result is called

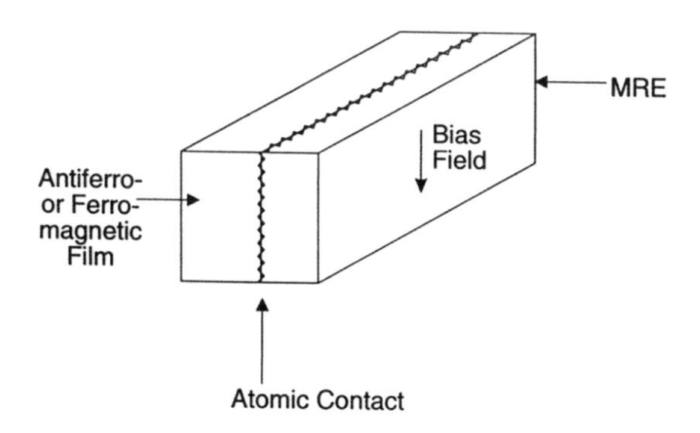

Fig. 5.6. The exchange bias principle.

exchange anisotropy. When $J_{EX} > Kt$, the effect is principally that of increasing the coercivity.

Antiferromagnetic films of MnFe, CoO, and TbCoFe and ferromagnetic films of CoP, CoNiP, and CoCrPt are used in MRH biasing. The ferromagnetic films may or may not be exchange coupled depending on whether atomic contact has been established. The exchange-coupling field of a ferromagnetic film is of opposite polarity to that film's magnetostatic coupling field.

The magnitude of the exchange field varies widely from a few oersteds to several hundred oersteds. The exact control of the exchange field is a very difficult materials science problem because it depends on atomic details of the interface that are, by their very nature, almost unmeasurable. The magnitude and sign of the exchange coupling between a pair of atoms is a rapidly varying function of the atom-to-atom spacing, as is shown in Fig. 1.4. The measured exchange field is the coupling averaged over millions of interfacial atoms.

The main advantage of exchange bias is that when an antiferromagnetic film is used, the bias field generated cannot be reset or changed accidentally during the lifetime of the MRH. To reset antiferromagnetic exchange anisotropy, it is necessary to cool the antiferromagnet from above its Néel temperature in the presence of a magnetic field. The Néel temperature is the antiferromagnetic analog of the ferromagnetic Curie temperature.

When a ferromagnetic exchange bias is used, the exchange bias can be reset or changed merely by using a magnetic field sufficiently greater than the coercivity of the ferromagnet.

A disadvantage of exchange bias, other than the difficult materials science and fabrication technology already discussed, is that many of the commonly used antiferromagnets are prone to oxidation and corrosion. The corrosion problem in antiferromagnetic films has led to their gradual replacement with ferromagnet films of high coercivity and anisotropy.

The Barber-Pole Scheme

In the barber-pole biasing scheme, an entirely different approach is taken to the fundamental vertical biasing problem of establishing the correct angle between M and I. As shown in Fig. 5.7, instead of using magnetic bias to rotate the MRE magnetization M to the appropriate angle θ, the current is made to flow obliquely in the MRE.

The permalloy MRE is overcoated with a thick conductor film that is photolithographically processed to etch out the "barber's pole" pattern shown. By making the conductor layer thick, essentially no current flows in the MRE in the regions below the conductors. Between the conductor stripes, however, the current must flow in the permalloy and it does so in the shortest possible path, that is, orthogonal to the conductor stripes. The result is that the proper angle between M and I is imposed by the geometry of the stripes.

In some designs, the stripes are arranged in a chevron pattern so that on one side of the MRE the M,I angle is positive and on the other negative. This elaboration is undertaken to reduce the asymmetry in the positive and negative amplitude response inherent in a single-element MRH. Pulse amplitude symmetry can also be obtained in the split-element and servo-bias schemes discussed next and in some of the double-element MRH designs discussed in Chapter 10.

When the barber-pole idea is simplified to its irreducible minimum, the M,I current angle θ can be established with merely a pair of "slant conductors." This technique has increasing appeal as the track width and MRE width W is decreased.

The main advantage of barber-pole bias is that the M,I angle is set directly. There are no complicated magneto-static (demagnetizing field or adjacent layer interaction field) complications.

Among the disadvantages is the unfortunate fact that parts of the MRE are, by design, not used to sense the signal flux from the recording medium. Estimates have been made that no more than about 60% of the MRE width W can be used as an active MR sensor.

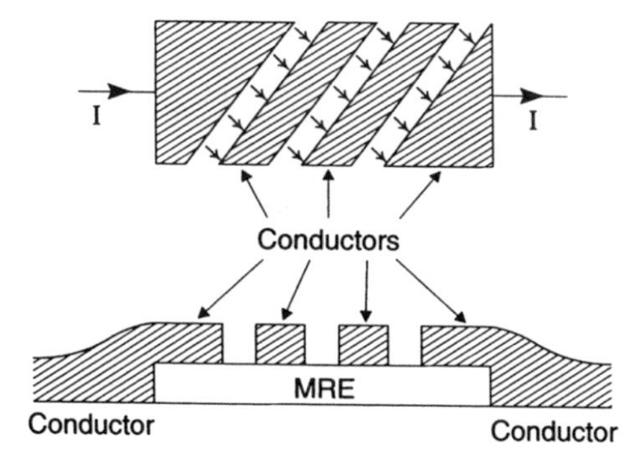

Fig. 5.7. The barber-pole structure.

The barber-pole vertical bias scheme is used in the Philips Digital Compact Cassette (DCC) audio recorder's yoke-type MRHs. Yoke-type MRHs are discussed in Chapter 9.

The Split-Element Scheme

The split-element idea, shown in Fig. 5.8, is not really a means of producing vertical bias fields. Rather, it is a clever method of minimizing the signal asymmetry or distortion that results when the vertical bias field, applied by one of the other methods, is not of the correct value.

By providing a center tap, the current in the MRE is made to flow in opposite directions on either side. This is shown in Fig. 5.8.

Suppose that a vertical bias field has been produced by one the methods using the MRE electrical current, for example, self-bias. Because the current flows in opposite directions in either side, the polarity of the vertical bias field is opposite in either side. In Fig. 5.8(A) the down vertical bias field is shown on the left-hand side and vice versa on the right.

Suppose that the magnitude of the vertical bias fields is not correct. In Fig. 5.8(B) the variable resistance versus signal field, H_{signal} characteristics of the two sides are shown with both being underbiased. To achieve small-signal linearity with minimal pulse amplitude asymmetry, the MREs require a greater magnitude of vertical bias field.

When the split-element senses the fringing field of the recording medium, H_{signal}, the changes of resistance are of opposite sense on either side. For the up signal field shown in Fig. 5.8(A), the resistance of the left side increases, whereas that of the right side decreases.

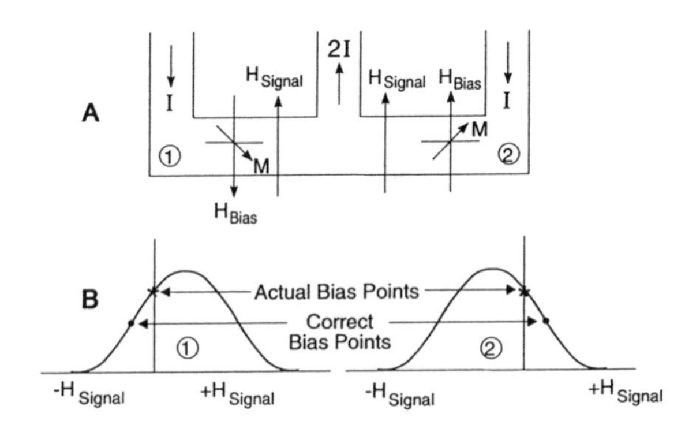

Fig. 5.8. (A) The split element arrangement and (B) the push–pull idea.

These resistance changes are sensed differentially. The differential sensing cancels out the pulse asymmetry due to the incorrect vertical biasing. In fact, the technique of differentially sensing the opposite polarity outputs of two devices is called "push–pull" operation in electronic design and it is extremely common in, for example, high-power audio amplifiers.

The advantage of the split-element idea is that the differential sensing improves the pulse amplitude symmetry due to improper vertical biasing, regardless of whether that bias is too small or too great. The scheme is, therefore, appealing in analog applications, such as audio recording, where the allowable (second) harmonic distortion is much lower than in the case in digital (binary) recording.

The main disadvantage is, apart from the circuit complexity, the fact that the portion of the MRE underneath the center conductor cannot sense signal because it has neither a bias field nor a measuring current. The split-element arrangement is used in IBM 0.5-in. tape drive MRHs.

The Servo-Bias Scheme

A very recent innovation in vertical bias technology is shown in Fig. 5.9. It is an evolution of the shunt bias scheme already discussed. In the servo-bias scheme, the current in the adjacent conductor must be electrically isolated from that in the MRE.

The current flowing in the adjacent conductor shown in Fig. 5.9 produces a down vertical bias field in the MRE. However, this current is not generated by a constant current (high-impedance) source, but rather from a variable current driver.

The variable driver is, in fact, a negative feedback electronic circuit whose output is the variable bias current. The error signal fed back to the bias current servo is derived from the magneto-resistive variable change, δR, of the MRE. The actual error signal could be, for instance, the total even harmonic distortion of the MRE voltage.

Thus the function of the bias current servo is to adjust automatically the bias current to the proper amplitude which gives the minimum even harmonic distortion. This value of the bias current produces precisely the correct vertical bias field in the MRE. It should be clearly understood that the precision which can be attained with automatic servoing of the vertical bias is limited only by the signal-to-noise ratio available in the negative feedback error signal.

Two modes of servo-bias operation can be visualized. In the first, the

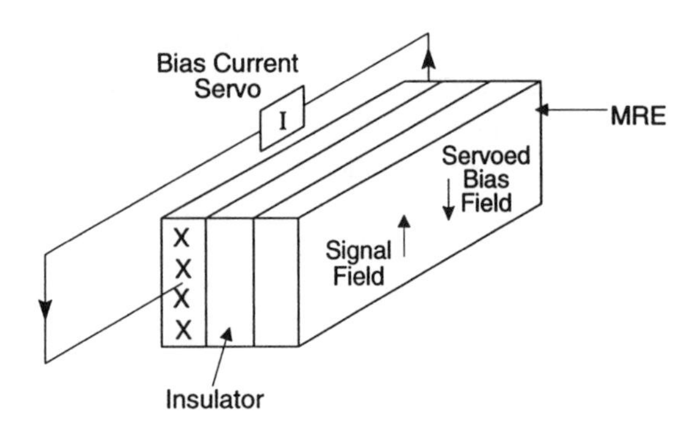

Fig. 5.9. The servo-bias system.

servo-bias current driver has only a low-frequency response or bandwidth. In this case the signal output is just the usual varying resistance of the MRE. In this mode, the servo-bias merely ensures that the MRE remains at the optimum bias point which minimizes, say, pulse amplitude asymmetry or even harmonic distortion.

In the second mode, the servo-bias driver has a much greater bandwidth than that of the recorded signals. Now the voltage drop across the MRE can be held fixed and the varying servo-bias current becomes the signal output of the MRH. In this case, subject only to the signal-to-noise ratio of the negative feedback error loop, the variable servo-bias current and output signal become almost free of harmonic distortion. Moreover, because the MRE is now simply a "null" detector, the actual slope of its variable resistance versus field characteristic is no longer of great importance!

The advantage of the servo-bias scheme is particularly obvious in multitrack MRHs where the problem of achieving uniformity in the resistance change versus current characteristic of each track is extremely difficult. The production yield of multitrack MRHs can be increased greatly by the larger track-to-track variations that are tolerable when servo-bias is used.

The disadvantages of servo-bias are equally obvious. A relatively complicated bias current servo circuit must be provided. The bias current servos currently used are implemented in a large-scale integrated (LSI) silicon chip.

Servo-bias is used in the Philips DCC audio recorder. In the latest version of the DCC, no less than 16 parallel tracks of MRHs are independently servo-biased! When reproducing analog tapes, the 16 servo outputs are summed as left and right stereo groups of 8.

As an aside, this writer cannot help noting that despite the fact that Bob Hunt invented MRHs at Ampex, no practical application for MRHs could be found in the analog audio, instrumentation, and video recorders made by Ampex in the period 1960–1990. The principal reasons for this failure to find applications were twofold. First, each individual head requires a slightly different vertical bias current for optimum operation. Second, even at the optimum point, the dynamic range (the range between maximum and minimum analog signal levels) was always too small, being limited by the harmonic distortion inherent in MRHs. Proposals made at Ampex, by this writer, to use a vertical servo-bias system were, in that time frame, deemed to be "obviously too complicated" or "too expensive" to implement. Now in the 1990s, no less than 16 parallel servo-bias MRHs are operated simultaneously in a portable consumer audio entertainment machine!

Further Reading

Additional reading material is listed here that will prove helpful to readers who seek more detailed information.

Bajorek, C. H., Krongelb, S., Romankiw, L. T., and Thompson, D. A. (1974), "An Integrated Magnetoresistive Read, Inductive Write High Density Recording Head," in *20th Annual AIP Conf Proc.*, American Institute of Physics, p. 24.

Kuijk, K. E., van Gestel, W. J., and Gorter, F. W. (1975), "The Barber Pole, A Linear Magnetoresistive Head," *IEEE Trans.* **MAG-11,** 5.

Markham, David, and Jeffers, Fred (1991), "Magnetoresistive Head Technology," *Proc. Electrochemical Soc.* **90,** 8.

Shelledy, F. B., and Nix, J. L. (1992), "Magnetoresistive Heads for Magnetic Tape and Disk Recording," *IEEE Trans.* **MAG-28,** 5.

CHAPTER

6

Horizontal Biasing Techniques

Horizontal biasing is necessary in MRHs so that the single-domain magnetization state of the MRE will be stable against all reasonable perturbations. These perturbations include external magnetic fields and thermal and mechanical stresses.

As indicated in Fig. 6.1, it is almost inevitable that the end zones of the MRE will not be in the single-domain magnetic state but that the magnetization will break up into the multidomain state. If the multidomain state did not occur in the end zones, the demagnetizing field in those zones would be extremely high ($H_d \approx 4\pi M_s \cos\theta$). In reality, of course, the demagnetizing field in the end zones cannot exceed the coercivity of the MRE because domain walls are nucleated and the multidomain state is created.

The primary function of the horizontal bias is to keep the end-zone domain walls pinned or fixed in position. Whenever domain walls move beyond a small reversible range, they jump irreversibly to new equilibrium positions. These irreversible domain wall jumps are the origin of hysteresis in soft, that is, low-coercive-force, magnetic materials.

When the horizontal bias is not adequate to pin the end-zone domain walls in position, domain wall jumps change the demagnetizing field in the MRE. This changes the single-domain magnetization versus current angle in the single-domain region and changes the variable resistance versus field characteristic curve. Because domain wall jumps show hysteresis, the resistance characteristic curve also displays hysteresis. Hysteresis in the resistance characteristic means that the increasing field curve is no longer the same as that for decreasing fields.

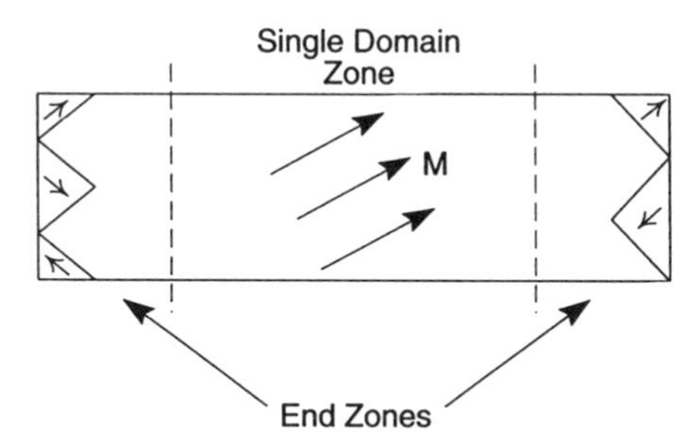

Fig. 6.1. The magnetization state of an MRE showing the single-domain center zone and the multidomain end zones.

Many of the techniques used for producing vertical bias have been proposed for horizontal bias. Again, the field is replete with novelty and ingenuity.

Clearly, the permanent magnetic (PM) film arrangement shown in Fig. 5.1, could be used for applying both vertical and horizontal bias fields by merely setting the PM film's magnetization at an oblique angle. Similarly, the exchange bias scheme shown in Fig. 5.6 could be used for both vertical and horizontal biasing. On the other hand, the vertical bias schemes that use electric currents cannot provide horizontal bias fields.

The horizontal biasing arrangements discussed in detail in this chapter are those which are not merely simple adaptations of vertical bias methods. Closely allied to horizontal bias techniques are other methods used to minimize the actual effect of unstable domain walls in the end zones. An outline of these methods is also included in this chapter.

Conductor Overcoat

Perhaps one of the simplest ideas is shown in Fig. 6.2. Here the MRE is simply made much wider than the required track width, W. The troublesome end zones of the MRE are arranged so that they do not carry any magneto-resistance measuring current. In Fig. 6.2 this is achieved by overcoating the end zones with a sufficiently thick conductor layer. By allowing no current to flow in the end zones, the end zones do not contribute to the MRH output signal.

Moreover, in the single-channel MRHs used in disk files, the end zones can be almost arbitrarily distant from the center current-bearing sensing region of the MRE and thus the effect of the end-zone domain wall instability can be made very small. A further elaboration on this idea

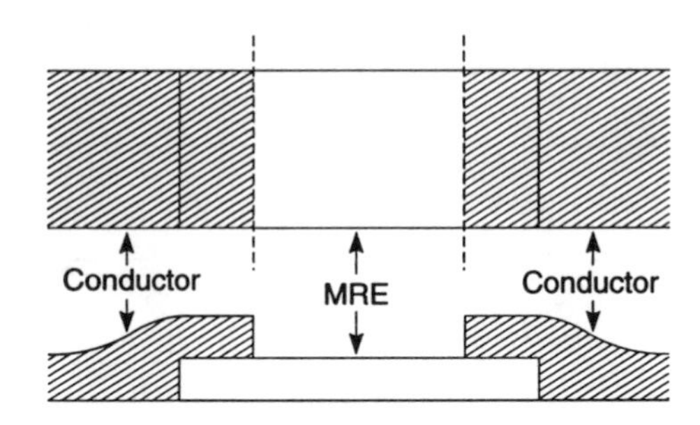

Fig. 6.2. The conductor overcoat technique.

is to progressively thin the MRE thickness T toward the end zones. This thinning raises the coercivity of the film and helps to pin the domain walls in place.

Another advantage of conductor track-width definition is that a precise MRE sensing track width W is set directly by the photolithographic process definition of the conductor overcoat. Furthermore, by careful design of the way the current leads join the conductor overcoat, useful horizontal bias fields can be generated in the end zones.

Exchange Tabs

Another approach to pinning the end-zone domain walls is shown in Fig. 6.3. Here, the end zones are exchange coupled to either a ferromagnetic or antiferromagnetic layer. Exchange coupling was discussed in detail in Chapter 5.

When the exchange coupling field is sufficiently large, the exchange tabs fulfill two roles. First, the end-zone domain walls are pinned and thus the MR characteristic displays no hysteresis. Second, even though measuring current is flowing through the MRE under the exchange tabs, no magneto-resistive signal is generated because the magnetization versus current angles cannot change in the pinned region.

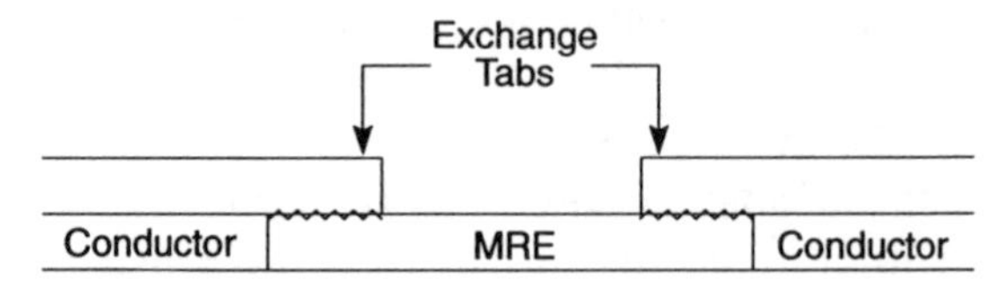

Fig. 6.3. The exchange tab arrangement.

Typically, exchange tabs made with MnFe have been used. This was the case in the first two generations of IBM disk file heads, which were code named "Sawmill" and "Corsair."

Hard Films

Figure 6.4 illustrates yet another approach to freezing the end-zone magnetization state. Films of high coercivity material are abutted against the ends of the MRE. We can understand how this works by thinking about it in two ways. First, by arranging for the flux flow ($B_s \cos\theta\, TD$) of the MRE to be equal to the flux flow of the hard film, we see that, because no magnetic poles exist at their junction, the demagnetizing field in the end zone vanishes. Alternatively, we can imagine that the magnetic poles at the ends of the hard film produce a sufficient horizontal bias field in the MRE.

In Fig. 6.4, the precise sensing track width of the MRH is defined by the current conductors.

An obvious practical difficulty in the implementation of hard-film horizontal bias arises. The efficacy of the scheme depends strongly on the microscopic details of the MRE–hard-film junction. This problem can be mitigated, but not eliminated, by sloping the interface so that the actual junction is spread over a distance as large as perhaps 0.1 μm.

Hard-film horizontal biasing has replaced exchange bias in later generation IBM hard-disk file heads, such as "Allicat," "Spitfire," "Starfire," and "Ultra."

Closed Flux

In multichannel MRH assemblies, use is made of yet another idea for reducing the effect of the end-zone domain wall instability. Figure 6.5 shows an MRE in which the end zones have been folded around above a conductor defined sensing region. The two end zones are separated by a small gap in order to prevent electrical shorting of the MRE. The entire structure shown is made of permalloy.

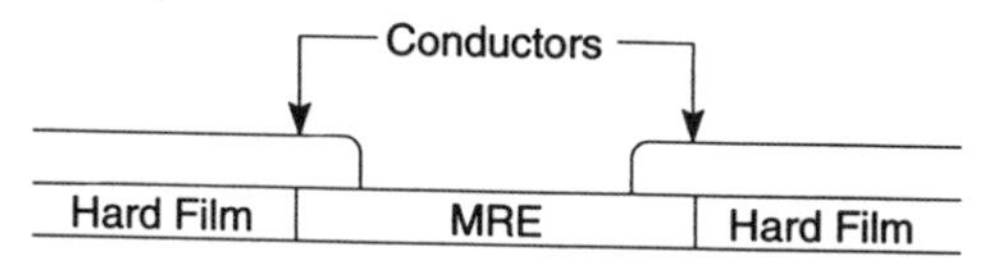

Fig. 6.4. The hard-film geometry.

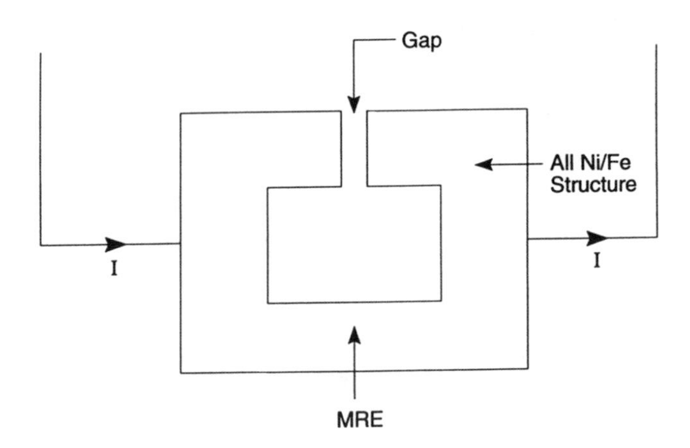

Fig. 6.5. The closed-flux structure.

Recall that a closed structure, such as a toroid, can be magnetized circumferentially without generating $\rho = -\nabla \cdot M$ magnetic poles and, therefore, both the internal (demagnetizing) or external (fringing) fields are zero. A divergence-free magnetization state is called a *closed-flux* pattern. The closed flux horizontal biasing idea is to create, as far as possible, such a divergence-free magnetization pattern in the MRE.

The closed-flux idea is particularly useful in multichannel MRHs where there is not sufficient intertrack spacing to apply the other horizontal biasing schemes discussed earlier in the chapter.

Closed-flux horizontal biasing is used in the Philips DCC in conjunction with barber-pole and servo-bias vertical biasing.

MRH Initialization Procedures

When the fabrication processes of an MRH have been completed, the device is usually subjected to initialization procedures. These procedures can be applied either to the thousands of batch-fabricated heads at the wafer level or to individual finished heads.

The purpose of the initialization processes is to set the proper magnetic state of the various parts of the MRH. Thereafter, the finished MRH has to remain in a reproducible stable magnetic state against all reasonable magnetic, thermal, and mechanical perturbations.

The parts which typically need to be initialized include those which produce the vertical and horizontal biasing and the MRE.

Initialization processes start with those that require high magnetic fields or high temperatures and finish with low-field, low-temperature steps.

When the device contains permanent magnetic materials, these are the first to be initialized. If the PM film has a coercivity, H_c, it is necessary to apply total fields of, typically, 3 to $4H_c$ in order to set the PM film at its maximum remanence. Here, *total field* refers to the vector sum of the external field and any demagnetizing or fringing fields generated by the magnetic layers in the MRH. Very often, external fields exceeding 10,000 Oe are necessary even with PM films that have an H_c of only 1000 Oe. This step sets the remanent magnetization of the hard film at the correct magnitude in the correct direction.

The next process, for devices which contain antiferromagnetic layers used for providing exchange bias, is that of setting the antiferromagnet. Just as ferromagnetic and ferrimagnetic materials have magneto-crystalline anisotropy, with easy and hard axes, so do antiferromagnetic materials. Setting the antiferromagnet means setting the (alternately opposite) spin directions on the correct easy axis of the antiferromagnet.

To initialize an antiferromagnet film, it must be cooled down through its Néel temperature in the presence of a magnetic field. The Néel temperature of an antiferromagnet is the analog of the Curie temperature of a ferromagnet. It is that temperature at which thermal energy ($\frac{1}{2}kT$) destroys the exchange-mediated antiferromagnetic ordering and the material becomes a mere paramagnet. The Néel temperatures of many of the antiferromagnets used in MRH technology are in the range of 300 to 400°C.

The magnetic field, which must be present during the cooling phase, need not be of large magnitude because, at the Néel temperature, the magneto-crystalline anisotropy energy K of the material vanishes. The field serves only the purpose of providing a preferred direction during the cooling. It is a magnetic process closely related to that of anhysteresis. Typically, field magnitudes of 100 Oe suffice and, of course, such low fields do not upset the PM film initialization which has already been done.

Finally, the magnetic state of the MRE has to be initialized. This process can have two parts. First, it is often necessary to perform a magnetic anneal in order to reestablish the easy axis of the MRE in the correct direction. The correct direction is usually the cross-track direction in both the usual anisotropic MRHs and in the giant MRHs discussed in Chapter 12.

Magnetic annealling in permalloy typically requires several hours at 250°C in a magnetic field. A total magnetic field strength sufficient to saturate the magnetization is required. Higher fields are not more effective. The field strength required is approximately 3 to $4H_c$ plus any demagnetizing and fringing field that exists. Typically, a field of several hundred Oersted suffices. Because the annealing temperature is less than the Néel temperature of any antiferromagnetic material in the structure, the exchange bias initialization is not perturbed.

The development of an induced anisotropy and easy axis in permalloy is called a *magnetization-induced* process and occurs by the development of an ordered atomic structure in which the 81Ni/19Fe spatial distribution of atoms is no longer statistically random.

The second MRE initialization process is often called "ringing down." The device is exposed to an alternating polarity sequence of magnetic fields which have decreasing magnitudes. The process is entirely analogous to the familiar process of "ac demagnetization," which is used to demagnetize everything from watches and recorded magnetic media to ships. Its purpose is to leave the MRE in the lowest attainable energy state, which, almost by definition, is most likely to have the greatest stability. The initial field magnitudes need only be sufficient to saturate the MRE magnetization. The process is usually performed at room temperature and, accordingly, it does not upset any of the previous initializations.

At the completion of these sequential processes, the MRH is expected to remain in the proper magnetic state throughout its operational lifetime.

Further Reading

Additional reading material is listed here that will prove helpful to readers who seek more detailed information.

Bajorek, C. H., Krongelb, S., Romankiw, L. T., and Thompson, D. A. (1974), "An Integrated Magnetoresistive Read, Inductive Write High Density Recording Head," in *20th Annual AIP Conf Proc.*, American Institute of Physics, p. 24.

Kuijk, K. E., van Gestel, W. J., and Gorter, F. W. (1975), "The Barber Pole, A Linear Magnetoresistive Head," *IEEE Trans.* **MAG-11,** 5.

Markham, David, and Jeffers, Fred (1991), "Magnetoresistive Head Technology," *Proc. Electrochemical Soc.* **90,** 8.

Shelledy, F. B., and Nix, J. L. (1992), "Magnetoresistive Heads for Magnetic Tape and Disk Recording," *IEEE Trans.* **MAG-28,** 5.

CHAPTER

7

Hunt's Unshielded Horizontal and Vertical Magneto-Resistive Heads (MRHs)

As stated in the preface, magneto-resistive heads (MRHs) were invented by Robert Hunt at Ampex in 1968. The first paper, written by Hunt, appeared in 1970. Hunt's paper deserves close scrutiny by all interested in MRHs because the principal advantages of using MRHs, rather than the usual inductive heads, are put so clearly. In the first paragraph, Hunt states:

> *The following advantages accrue to a magnetoresistive transducer (MRT). 1) The device measures the field rather than the time derivative of flux, making output levels insensitive to scan velocity. 2) The device has no intrinsic frequency limitations for practical bandwidths (i.e., <100 MHz). 3) The device has no intrinsic noise mechanisms to deteriorate signal-to-noise ratio. 4) Many device elements may be mounted side by side on the same substrate and registered to within optical tolerances, eliminating the serious problem of gap scatter in ring head technology. 5) The transducer output levels are characteristically in the 10–100 mV (rms) range. 6) The MRT's wavelength characteristics, except for a gap interference term, are similar to ring head response. A disadvantage of the device lies in its inability to record signals onto tape.*

Point 1, that the output voltage is independent of scan velocity, is the principal reason for the widespread application of MRHs in digital tape and small-diameter hard-disk drives. This fact is expanded upon in Chapter 11.

Point 2 is as true today as it was in 1970. Magnetization reversal processes, which involve only rotation of the magnetization in a single domain, are several orders of magnitude faster than those that proceed by domain wall motion. It is indeed notable that device operation at 100 MHz, capable of supporting some 200 Mbits/second, was being investigated experimentally at Ampex as early as 1968. Such data rates, though of great interest in digital video recording, have not yet been attained in computer peripheral recorders even 25 years later.

Today, point 3 is understood to mean that, when appropriate horizontal biasing is applied in order to suppress domain wall jump in the MRE end zones, the MRH noise power is simply the Johnson, or thermal noise ($4kTR\Delta f$), of the MRE. Here, k is Boltzmann's constant, T is the absolute temperature, R is the resistance of the MRE, and Δf is the noise measurement bandwidth.

Point 4 is precisely the attribute of thin-film MRHs that is exploited in the 0.5-in. tape IBM digital recorders (IBM 3480 and 3490), which have 18 and 36 parallel tracks, respectively. Similarly, the Philips DCC compressed digital audio consumer machines have 16 parallel tracks across a 0.15-in.-wide (4-mm-wide) tape.

Point 5, that output voltages as high as 100 mV rms (over a 50-mil track width) can be realized, is within a factor of two of the voltages attained today in permalloy anisotropic MRH heads. We will see in Chapter 14 that the same specific output, that is, the voltage per unit track width, should always be expected from *any correctly designed* anisotropic MRH with a permalloy sensor. It is a property of the permalloy only.

Point 6, concerning the MRT's wavelength characteristics or spatial frequency spectrum, is discussed later in this chapter.

Finally, Hunt points out that MRHs cannot record signals and this is, of course, absolutely correct. It follows that a separate write head must be used with an MRT or MRH. Recall, however, that the physical processes of recording and reading are entirely different, as was reviewed in Chapters 2 and 3. For example, the optimum writing gap is invariably longer than that needed for optimum reading.

When a separate write head is used, it becomes possible to accomplish several other desirable changes. For example, the number of turns on the dedicated write head can be reduced, because that number is usually set by reading criteria in dual-role inductive write–read heads. Reducing the number of turns in write heads not only simplifies the head fabrication process, but also reduces the voltage compliance required of the write current driver because the head impedance is reduced. Moreover, the separation of writing and reading heads into separate structures makes possible the strategy of "write wide–read narrow" in which the read channel signal (and signal-to-noise ratio) is deliberately compromised, in order to alleviate track-following problems during read operations.

We can conclude that although Hunt was absolutely correct about the disadvantage of an MRT or MRH, we now realize that many advantages accrue with the separation of write and read heads that is mandated by MRHs.

The Unshielded Horizontal Head

Hunt's first MRH is shown in Fig. 7.1. Here the permalloy sensor is laid flat across the recording tape. The MRE senses the medium fringing field component that is in the plane of the thin film. This is the horizontal or x component of the fringing field of the medium. The MRE dimensions used were $W = 50$ mils, $D = 1$ mil, and $T = 1000$ Å.

The horizontal field, measured at a distance y above a tape sinusoidally magnetized uniformly throughout its depth, is known. Let the magnetization waveform $M(x)$ recorded be

$$M(x) = M_R e^{jkx}, \tag{7.1}$$

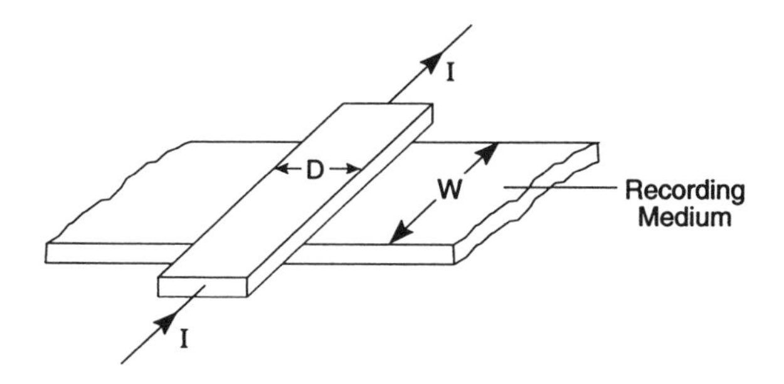

Fig. 7.1. Hunt's horizontal MRH.

where M_R is the maximum remanence of the tape or disk, $j = \sqrt{-1}$, k is the wavenumber ($=2\pi/\lambda$), and λ is the wavelength. The exponential term e^{jkx} is called the *generalized sinusoid* and is equal to $(\cos kx + j \sin kx)$.

The horizontal component of the fringing field above the tape is

$$H_x(x) = -2\pi M_R(1 - e^{-kd})\, e^{-ky}\, e^{jkx}, \tag{7.2}$$

where δ is the tape thickness and y is the distance above the tape.

The term in parentheses is called the *thickness loss* and shows that 90% of the maximum possible field H_x comes from a surface layer in the medium only about one-third (0.36 actually) of a wavelength deep.

The exponential term e^{-ky} is called the *spacing loss* and it shows that the field H_x falls to $e^{-2\pi}$ (1/535th) of its value on the surface of the medium when $y = \lambda$. In decibels, the spacing loss is $-54.6y/\lambda$ dB.

We can see from Eq. (7.1) that the magnitude of the horizontal field H_x changes across the element dimension D. Accordingly, the magneto-resistance changes across the dimension D. A simple way to visualize the effect of these changes is to imagine the MRE divided into a number N of strips parallel to the W dimension and the current flow and to consider these strips as a set of variable resistors connected in parallel. Provided the MR coefficient $\Delta\rho/\rho$ is small, it can be shown by Ohm's law that, if each strip has resistance $R + \delta R$, the overall change in resistance of the parallel set is just the average of the changes, $\Sigma\delta R/N$. Of course, the MRE is continuous and the overall chance in resistance is actually given by $1/D \int \delta R(x)\, dx$. Here $\delta R(x) = H_x(x)\, dR/dH$, where dR/dH is the slope of the variable resistance versus field characteristic.

Following the straight-line change in resistance versus field approximation of the biased MRE characteristic that was discussed in Chapter 4, the sensitivity or slope is $-\frac{1}{2}\Delta R/H_{\text{bias}}$. The change in resistance of a single strip is, accordingly, proportional to $\frac{1}{2}\Delta R H_x/H_{\text{bias}}$.

When these changes of resistance are averaged over the MRE dimension D and multiplied by the measuring current I, the result is the output voltage spectrum for small signals ($H_x < H_{\text{bias}}$),

$$V_H(k) = \left(\frac{I\Delta R}{2}\right) \left[\frac{2\pi M_R(1 - e^{-k\delta})\, e^{-kd}}{H_{\text{bias}}}\right] \left(\frac{\sin kD/2}{kD/2}\right), \tag{7.3}$$

where d is the MRE-to-tape spacing.

In this expression, the first bracketed term on the right side, the measuring current times one-half the maximum change in MRE resistance, is obviously a voltage. The other bracketed terms are dimensionless.

The second term is the magnitude of the tape field, H_x, divided by the bias field, H_{bias}. Again, it shows the thickness loss and spacing loss terms common to all read heads that sense the fringing field or flux above a recorded medium.

The third bracketed term is a direct result of the averaging over the distance D discussed earlier. It is precisely the "gap interference" term mentioned by Hunt.

Note that Eq. (7.3) expresses the output voltage spectrum. The spectrum is the magnitude of the output signal at a particular wavenumber, k $(=2\pi/\lambda)$. The actual output voltage waveform is, of course, $V_H(x) = V_H(k)\, e^{jkx}$. Because the MRH response has been made linear, by using the proper bias field H_{bias}, the output for an input generalized sinusoid at a single spatial frequency k contains only that fundamental frequency. Nonlinearities introduce higher harmonics, $2k$, $3k$, etc.

Hunt realized that his horizontal head was not satisfactory because the "gap interference" term severely limits the short-wavelength response. For example, at a wavelength $\lambda = D$, the response is zero. This behavior is analogous to that of an inductive head of gap length g, where the first gap null occurs at $\lambda = g$. The solution to this problem is to turn the MRE on end so that now the "gap interference" term becomes $(\sin kT/2)/kT/2$. Since T is the thin-film thickness, the first "gap interference" null now occurs at inaccessibly short wavelengths.

The Unshielded Vertical Head

Hunt's vertical head is shown in Fig. 7.2. Again, it senses only the component of the medium fringing field that is in the plane of the MRE. This is the vertical, or y, component of the field.

The vertical field, H_y, of the sinusoidally written medium is

$$H_y(x) = j2\pi M_R(1 - e^{-k\delta})\, e^{-ky}\, e^{jkx}. \tag{7.4}$$

The appearance of an extra factor, j $(=\sqrt{-1})$, missing in the analogous expression for the horizontal field H_x, should be noted carefully. It indicates that the vertical field H_y lags 90° out of phase with the written magnetization. If a sine wave is written, the vertical field follows a cosine wave because $\sin(\theta + 90°) = \cos\theta$.

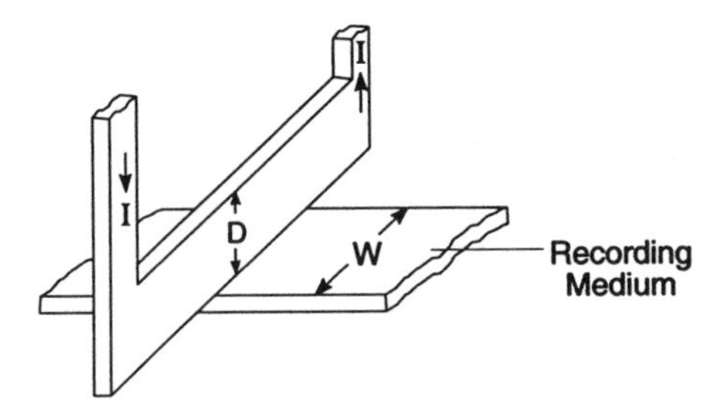

Fig. 7.2. Hunt's vertical MRH.

This 90° phase shift is of great importance because a normal inductive reading head also displays a 90° phase shift due to the Faraday law differentiation ($V = -N\, d\phi/dt$) of the head flux. It follows that the digital pulse shape and symmetry of a vertical MRH must be similar and identical, respectively, to that of a normal inductive head.

Different depths of the vertical MRH sense different magnitudes of vertical field H_y. Accordingly, the differing resistance changes over the depth have to be averaged.

The output spectrum for the vertical head is, for small signals ($H_x < H_{\text{bias}}$),

$$V_V(k) = j\left(\frac{I\Delta R}{2}\right)\left[\frac{2\pi M_R(1 - e^{-k\delta})\, e^{-kd}}{H_{\text{bias}}}\right] \left(\frac{1 - e^{-kd}}{kd}\right)\left(\frac{\sin kT/2}{kT/2}\right). \tag{7.5}$$

The significance of the factor j and first two terms on the right side has already been discussed. The third term is the result of the averaging of the small resistance changes δR over the depth D of the MRE. At long wavelengths, this term is approximately equal to unity and is thus not an important factor. However, at short wavelengths, $\lambda \leq D$, the term becomes approximately equal to $1/kD$. Thus, the vertical head output voltage falls off at short wavelengths at 6 dB/octave (20 dB/decade).

A simplistic way to think about this severe short-wavelength loss is to realize that the vertical field H_y, sensed by the parts of the MRE most distant from the medium, is very small because it falls off as e^{-ky}. The distant regions of the MRE thus act merely to "shunt out" electrically the resistance changes in the parts of the MRE that are close to the medium.

Despite this severe short-wavelength loss mechanism, in many applications, the extremely high output levels and the simplicity of construction of the vertical MRH, compared with that of inductive heads, make them the preferred transducer. Vertical heads, which have more recently been called simply unshielded magneto-resistive heads or UMRs, are used in many bank and credit card readers.

Single Magnetization Transition Pulse Shapes

When considering the performance of both MRHs and inductive heads for applications in digital (or binary) magnetic recording, it is often necessary to consider the so-called "isolated" output pulse shape rather than the output voltage spectrum.

Provided the response of the reproduce head can be considered linear, the pulse shape and the spectrum are, of course, closely related by simple Fourier transforms.

Suppose the written magnetization transition has the arctangent form $M(x) = 2M_R/\pi \tan^{-1}(x/f)$, which was discussed in Chapter 2. The Fourier transform of this magnetization transition $M(k)$, is equal to $2M_R e^{-jkf}/jk$ and the output pulse spectrum is simply $M(k)V(k)$. The pulse is merely the inverse Fourier transform of $M(k)V(k)$.

The advantages of frequency-domain analysis are twofold. First, sequential physical operations result in the simple multiplication of frequency-dependent terms, rather than a tiresome sequence of convolution integrals within convolution integrals. Second, in the frequency-dependent terms the physical variables have been "separated." Thus the spectra discussed earlier are the product of terms involving just the medium thickness δ, just the head-to-medium spacing d, just the MRE depth D, and just the element thickness T. A similar separation of variables does not occur when a linear system is described in the spatial or x domain.

Nevertheless, it is instructive to compare the isolated pulse shapes obtained with the vertical MRH and an inductive head.

Consider an inductive head with an almost "zero gap length," which means the gap length g is small compared with the other dimensions (f, δ, d) of the problem.

The isolated output pulse shape is

$$V_R(x) \propto \log \frac{(d + \delta + f)^2 + x^2}{(d + \delta)^2 + x^2}. \tag{7.6}$$

In the mathematical limit of sharp magnetization transitions ($f = 0$) and thin media ($\delta \rightarrow 0$), this reduces to the familiar Lorentzian pulse shape, $2\delta d/(d^2 + x^2)$, so frequently assumed in signal processing studies.

The 50% pulse amplitude width, PW_{50}, with the "zero-gap" inductive head is

$$_{R}PW_{50} = 2[(d + f)(d + \delta + a)]^{\frac{1}{2}}, \tag{7.7}$$

which, in the limit of $f \rightarrow 0$ and $\delta \rightarrow 0$, is just $PW_{50} = 2d$. Thus, the smallest imaginable PW_{50} for an inductive head is just twice the read head-to-medium spacing d. This is an interesting piece of knowledge in its own right.

For the unshielded vertical head, a similar analysis for the sharp transition, thin medium case yields,

$$V_V(x) \propto \log \frac{(d + D)^2 + x^2}{d^2 + x^2}, \tag{7.8}$$

$$_{V}PW_{50} = 2[d(d + D)]^{\frac{1}{2}}. \tag{7.9}$$

The preceding expression for the pulse width shows very clearly the undesirable effect of the electrical "shunting" or resistance change averaging over the MRE element depth D in the unshielded vertical MRH. The greater the MRE depth D, the greater the PW_{50}. Conversely, note that if D, the MRE depth, could be made negligible ($D < d$), then the limiting pulse width becomes $2d$, just as in the inductive head case.

The solution of this fundamental magnetization transition resolution or isolated pulse width problem in vertical MRHs is discussed in the next chapter.

Side-Reading Asymmetry

Consider the response of an MRE to the off-track magnetic poles shown in Fig. 7.3. When these poles are off-track to the left, the fringing field H_1 is almost orthogonal to the MRE magnetization. Accordingly, H_1 is able to rotate the magnetization by a relatively large angle, because the torque $M \times H_1$ is large.

On the other hand, the fringing field H_2, from magnetic poles off-track to the right, is almost antiparallel to the MRE magnetization. Now the torque is almost zero and the MRE's response is relatively small.

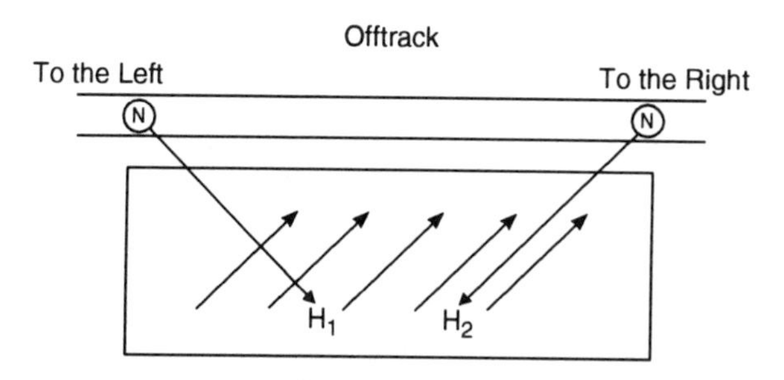

Fig. 7.3. The off-track, or side-reading, asymmetry of a single-element MRH.

Side-reading, or off-track asymmetry, is inherent in single-element MRHs. It can be reduced, but not eliminated, by using the barber-pole biasing technique. Perfect off-track symmetry can be achieved in some double-element MRH designs as is discussed in Chapter 10.

Further Reading

Additional reading material is listed here that will prove helpful to readers who seek more detailed information.

Hunt, Robert P. (1970), "A Magneto-resistive Readout Transducer" (digest only), *IEEE Trans.* **MAG-6,** 3.

Hunt, Robert P. (1971), "A Magneto-resistive Readout Transducer," *IEEE Trans.* **MAG-7,** 1.

CHAPTER

8

Single-Element Shielded Vertical Magneto-Resistive Heads

The shielded vertical magneto-resistive head (MRH) design overcomes the short-wavelength "shunting" deficiencies of the unshielded MRH discussed in the previous chapter. The basic design, shown in Fig. 8.1, is simply that of placing a vertical magneto-resistive element (MRE) in the gap between two shields made of a magnetic material of high permeability. To avoid confusion, the space between the MRE and a shield is referred to as the *half-gap* throughout this book.

In Fig. 8.1, only the MRE is shown. Note, however, that an actual head will include the extra structures necessary to provide both the proper vertical and horizontal bias discussed in Chapters 5 and 6.

The Function of the Shields

The function of the high-permeability shields can be understood by comparing Figs. 8.2 and 8.3, which show an unshielded and shielded vertical MRE, respectively. In the unshielded configuration, the vertical component, H_y, of the tape or disk fringing field can be large even when the sources of the fringing field, the $\rho = -\nabla \cdot M$ magnetic poles, are far away from the MRE. This "distant sensing" of the unshielded MRH causes its undesirable wide output pulse and short-wavelength spectral roll-off behavior.

When shields are added, however, the magnetic flux flow pattern

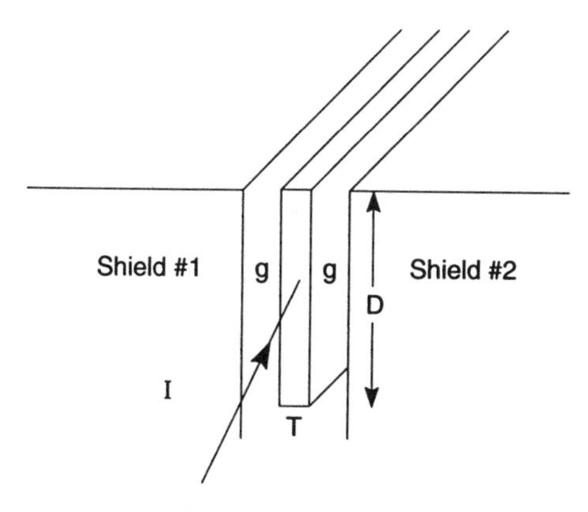

Fig. 8.1. Perspective diagram of a shielded, single-element MRH.

around the recorded medium is changed profoundly. The changes are due to the changed boundary conditions imposed by the shields. On the surface of any magnetic material, the exact boundary conditions are

$$B_{\text{normal}}\ (\text{outside}) = B_{\text{normal}}\ (\text{inside}), \tag{8.1a}$$

$$H_{\text{tangential}}\ (\text{outside}) = H_{\text{tangential}}\ (\text{inside}). \tag{8.1b}$$

When the shield has a very high permeability, the second condition becomes simply $H_{\text{tangential}}$ is equal to zero. As shown in Fig. 8.3, the flux from the recorded medium is attracted to the shields (the "keeper" effect) and it has to enter the top of the shields at right angles.

By making the shields extend well below the MRE depth D, almost

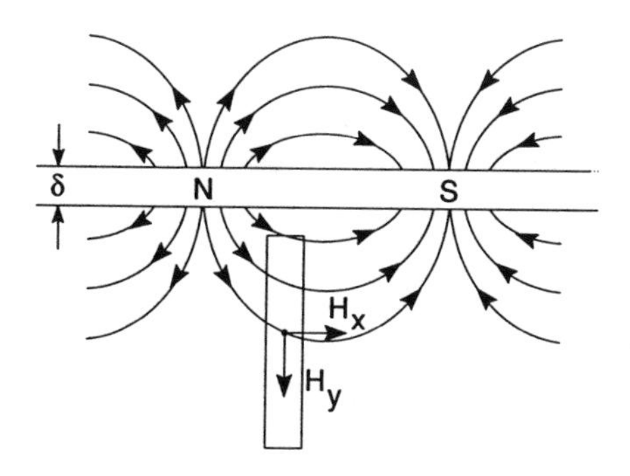

Fig. 8.2. The field and flux flow around an unshielded vertical MRH.

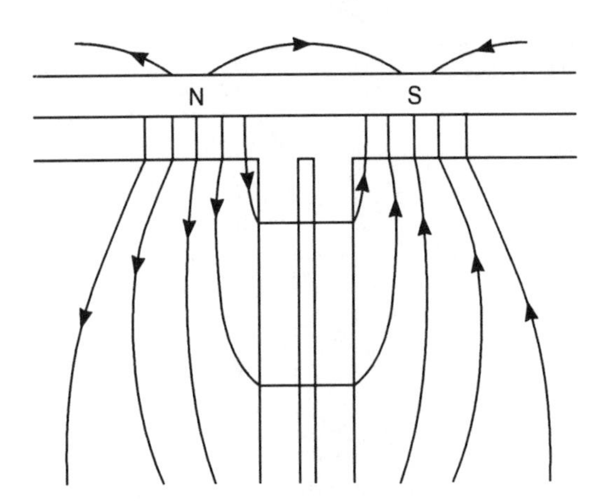

Fig. 8.3. The flux flow through the MRE and shields, when the magnetic transitions are not over the gap.

none of the media flux flows through the MRE, as is shown in Fig. 8.3. In this way, the MRE is shielded from the effects of distant poles.

When, however, the tape or disk is positioned so that a magnetic pole-bearing region is almost directly over the MRE, the flux pattern changes to that shown in Fig. 8.4. Here magnetic flux from the north pole shown flows directly into the top of the MRE. As the flux flows down the MRE, it leaks off sideways through the half-gaps into the shields and returns to the distant south poles. The MRE shield system is the magnetic analog of an electrical transmission line.

The overall effect of the shields is, therefore, to defer appreciable flux in the MRE until the last possible moment as the poles actually cross the top of the MRE. Recalling that it is the "line-of-sight" magnetic poles that, to first order, always produce the greatest magnetic fields and fluxes, a simple analogy of the shielded MRH's behavior can be made. A pedestrian is only able to see an overlying aircraft as it actually passes directly over an urban canyon, such as Wall Street in New York City, created by adjacent skyscrapers. Similarly, the MRE only "sees" the medium's magnetic poles as they actually pass over the top of the MRE.

To calculate the output voltage spectrum of the shielded MRH, it is necessary to understand not only the magnetic transmission line phenomenon but also the dependence of the MRE flux on the recorded wavenumber k.

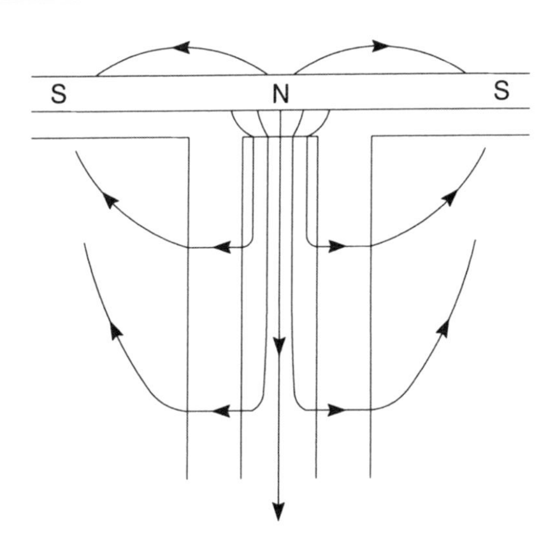

Fig. 8.4. The flux flow through the MRE and shields, when a magnetic transition is over the gap.

The MRE Shield Magnetic Transmission Line

Just as electric charge or current leaks from the center conductor to the outer shield of a coaxial cable, the magnetic flux entering the top of the MRE leaks into the shields.

Consider first the case where the MRE depth is very large as depicted in Fig. 8.5. It can be shown that the flux, $\phi(y)$, flowing down the MRE is,

$$\phi(y) = \phi(0)\, e^{-y/l}, \tag{8.2}$$

where $\phi(0)$ is the flux entering the top of the MRE and l is the transmission line decay length.

The decay length is the distance required for the side leakage to reduce the MRE flux to $1/e$ (0.367) of $\phi(0)$. The decay length is

$$l = \left(\frac{Tg\mu}{2}\right)^{\frac{1}{2}}, \tag{8.3}$$

where μ is the permeability of the MRE. The lower the MRE magnetic reluctance (larger T and μ), the longer the decay length. The higher the side-leakage reluctance (larger g), the longer the decay length. The factor of two arises because the side leakage occurs into both shields.

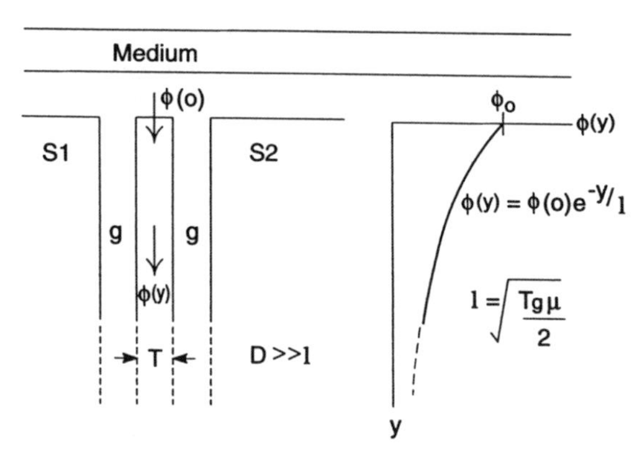

Fig. 8.5. The magnetic transmission line behavior of the MRE and shields.

Now consider the case where MRE depth D is less than or comparable to decay length l. This case is shown in Fig. 8.6. Now the MRE flux must become zero at the lower end of the MRE. The situation is analogous to that of an open-circuited electric transmission line, where the open circuit is usually treated by supposing that it is the source of a 180° out-of-phase current wave. Recalling that current in an electric circuit corresponds to flux in the equivalent magnetic circuit, we see that the lower end of the MRE can be considered to be the source of a negative (i.e., upward-flowing) flux.

The net result, shown in Fig. 8.6, is that the MRE flux, $\phi(y)$, de-

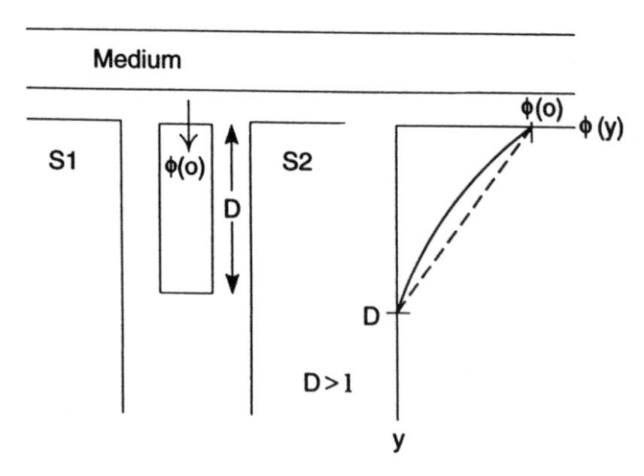

Fig. 8.6. The linear decay of flux with depth when the MRE depth is comparable to the magnetic transmission line characteristic length.

creases almost linearly from the top to the bottom of the MRE. The decrease is at an almost constant slope, when $D \leq l$, because neither the downgoing nor the upgoing flux depart much from constant slope.

In Chapter 7, we saw that the overall change in resistance of an MRE that is subject to a nonuniform field or flux is proportional to the average flux in the MRE. In the shielded MRH with $D \leq l$, the average is obviously always close to 0.5 $\phi(0)$ regardless of the actual depth D.

This observation is important in three ways. First it shows that the effective efficiency of the shielded MRH is close to 50%. The efficiency is defined in a similar way to that in which the efficiency of inductive read heads is defined. The inductive read-head efficiency is that fraction of the flux entering the top of the head that actually threads the coil.

Second, the fabrication of shielded MRHs is facilitated because the dimension D can have wide tolerances.

Of greatest importance, however, is that, because the MRE thickness T must be chosen so that the top of the MRE does not become magnetically saturated [$\phi(0) \leq B_s TW$], all optimally designed single permalloy element shielded MRHs produce approximately *the same maximum specific output voltage*. The maximum specific output voltage is about 2 V peak per centimeter of track width and it is discussed again in Chapters 11 and 14.

In some shielded MRH designs, the MRE is not the only highly permeable film between the shields. For example, when the soft adjacent layer (SAL) vertical bias technique is used, the second film has almost the same thickness–permeability product as the MRE. In such designs, because approximately 50% of the flux from the medium flows in the MRE and 50% in the SAL, the optimum design requires that thinner films be used. To keep the top of the MRE and the SAL just below magnetic saturation, their thicknesses are usually reduced to about one-half. It follows that decay the length l is almost unchanged. Moreover, the efficiency of the MRE remains at 50% and, therefore, the maximum specific output voltage is unchanged.

If the MRE is not electrically isolated and is shunted by, for example, a current bias or soft adjacent layer, the maximum specific output voltage is reduced accordingly. When the MRE and parallel shunting layer resistances are R_M and R_S respectively, the maximum specific output voltage becomes $2R_S/R_S + R_M$.

This behavior of the flux $\phi(y)$ in the MRE contrasts sharply with its behavior in the unshielded vertical MRH. The unshielded MRH, of

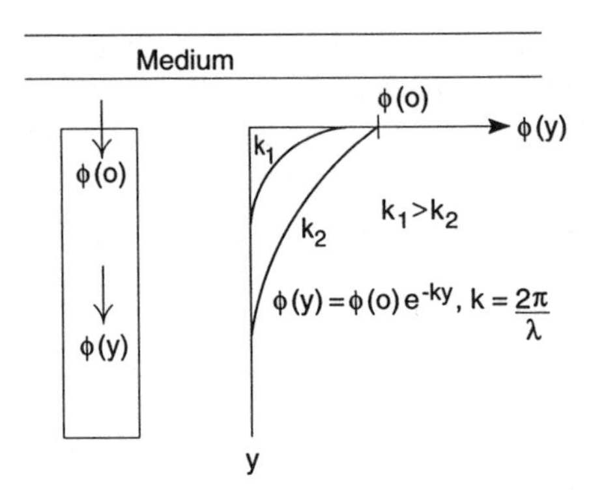

Fig. 8.7. The wavenumber-dependent exponential decay of flux with depth in an unshielded MRE.

course, does not have the massive shields of high-permeability material that alter the fringing flux pattern of the medium. Because the MRE is so thin (small T), we can assume that it has almost no effect on the medium's fringing flux pattern. Accordingly, the flux flowing down the MRE, $\phi(y)$, is

$$\phi(y) = \phi(0)\, e^{-ky}. \tag{8.4}$$

As indicated in Fig. 8.7, the fall-off of flux now depends on the recorded wavelength. The fall-off is just the spacing loss discussed earlier in this book.

The essential difference between an unshielded and a shielded MRH is that the decay of flux in the latter is independent of the recorded wavelength.

The Shielded MRH Output Voltage Spectrum

The dependence of the flux, $\phi(0)$, entering the top of the MRE can be elucidated in an almost nonmathematical manner by an extremely elegant application of the reciprocity theorem. The reciprocity theorem states that all (linear) reading heads which, when energized with current, produce exactly the same fringing field in the tape or disk must produce exactly the same output flux and voltage spectra and pulse shapes.

Consider Fig. 8.8, which shows the flux flow pattern caused by current flowing in a small coil wrapped around the MRE of a shielded

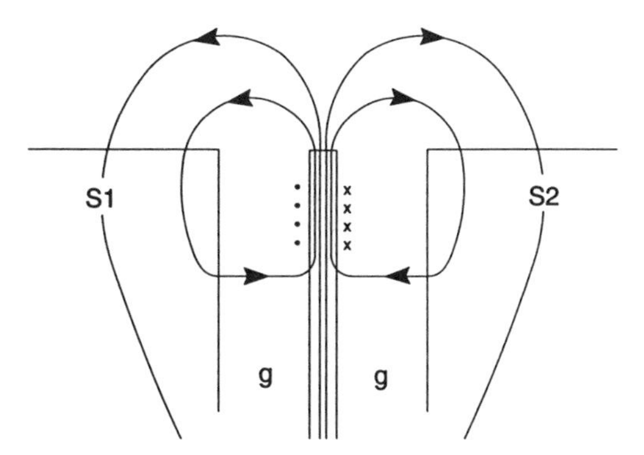

Fig. 8.8. The reciprocity field of a shielded MRH excited by a small coil wrapped around the MRE.

MRH. This coil does not exist in real shielded MRHs. It appears in Fig. 8.8 solely for the purposes of the reciprocity argument. Above the head, the fringing flux flows to the left and right over the left and right half-gaps, respectively. This flux pattern can be decomposed into two parts as indicated in Fig. 8.9. The fringing flux pattern above the left half-gap is almost the same as that above the gap of a normal ring head (#1)

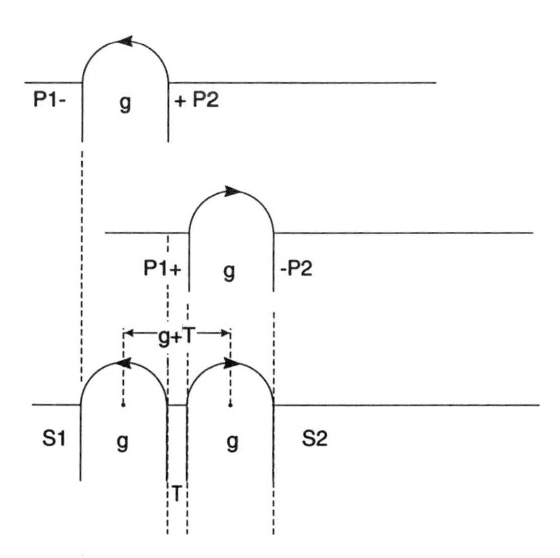

Fig. 8.9. The decomposition of the reciprocity field into the fields of two, oppositely polarized, and horizontally displaced ring heads.

polarized as shown. Similarly, the flux above the right half-gap is like that of another ring head (#2) of opposite polarity.

The fringing flux above a shielded MRH, excited by a small coil, is equivalent to that of two normal ring heads of opposite polarity which are offset from each other by the distance $(g + T)$. By the reciprocity theorem, we can conclude that the flux entering the top of the MRE, $\phi(0)$, must be equal to the difference in the fluxes of the two offset ring heads. Note that this simple conclusion has been arrived at directly, without recourse to mathematical analysis.

The flux flow around a ring head was discussed in Chapter 2. It is, again,

$$\phi_R(k) = -4\pi M_R W \left(\frac{1 - e^{-k\delta}}{k} \right) e^{-kd} \frac{\sin kg/2}{kg/2}, \qquad (8.5)$$

where g is the ring head whole gap length. The minus term simply means that the flux is 180° out of phase with the written magnetization pattern in the medium.

The MRE flux, $\phi(0)$, is just the difference in the flux of two ring heads that are offset by $(g + T)$. It is convenient to consider the offset of the two heads to be symmetrically $\pm(g + T)/2$. In the spatial frequency domain, these offsets are equivalent to multiplying by $\exp[\pm jk(g + T)/2]$ since the offsets introduce phase shifts of magnitude $\pm k(g + T)/2$ rad.

Recalling that $(e^{jx} - e^{-jx})/2j$ equals $\sin x$, the MRE flux $\phi(0,k)$ can be written as

$$\phi(0,k) = \phi_R(k) 2j \sin k(g + T)/2. \qquad (8.7)$$

Finally, the small-signal $(\phi_R < H_{\text{bias}} TW)$ output voltage spectrum of the shielded MRH with $D \leq 1$, can be written, by inspection, as

$$V_s(k) = \frac{1}{2} \left(\frac{I \Delta R}{2} \right) \left[\frac{4\pi M_R (1 - e^{-k\delta})\, e^{-kd}}{k} \right] \left(\frac{1}{H_{\text{bias}} T} \right) \times \left(\frac{\sin kg/2}{kg/2} \right) 2j \left(\frac{\sin k(g + T)}{2} \right). \qquad (8.8)$$

The enclosed terms on the right side on this spectrum can be identified with specific physical phenomena. The first term is the 50% efficiency of the head. The second is one-half the maximum resistance change of the MRE times the measuring current and is obviously a voltage. The third term is the flux per unit track width in a small gaplength ring head.

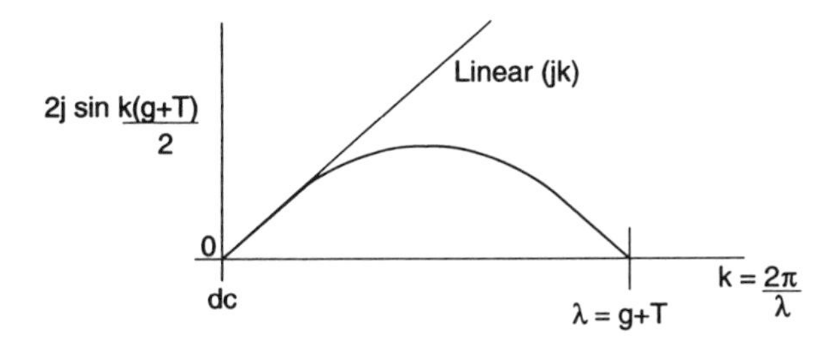

Fig. 8.10. A plot of the ring head displacement function versus wavenumber.

The fourth term is the bias flux per unit track width that has been produced in the MRE by the vertical bias field. The fifth term is the usual Karlquist approximation of the gap-loss function of a ring head with whole gap length, g.

The last term, a result of the flux subtraction between the offset ring heads, is plotted in Fig. 8.10. Note the factor j, which shows that the output spectrum suffers a 90° phase error, just as does the output voltage spectrum of an inductive head. At the short-wavelength $\lambda = (g + T)$ the shielded MRH has zero output. At medium to short wavelengths, say, $\lambda < 2(g + T)$, the shielded MRH effectively has two "gap-like interference terms." One is due to the half-gap g and the other to their separation $(g + T)$.

For many purposes, a convenient approximation is to pretend that a shielded MRH, with half-gap length g has the same spectrum as does a ring head with a whole gap length equal to $1.6g$.

At long wavelengths, the differencing term becomes approximately equal to $2jk(g + T)/2$. The Fourier transform of differentiation, $d(\cdot)/dx$, is jk, and this is also shown in Fig. 8.10. We see, therefore, that the differencing term is an approximation of differentiation. This fact will not surprise those readers who are involved in signal processing, because the standard circuit implementation of differentiation consists merely of subtracting a delayed version of a signal from itself.

The fact that the shielded MRH acts approximately as a differentiator has profound consequences. The normal Faraday's law inductive head output voltage is proportional to $-N\, d\phi/dt$, which is equal to $-NV\, d\phi/dx$, where N and V $(=dx/dt)$ are the number of turns and the head-to-medium relative velocity, respectively. The fact that the shielded MRH output voltage is also proportional to $d\phi/dx$ means that both its spectral shape and pulse output shape closely match those of inductive heads.

The shielded MRH is, therefore, effectively "plug replaceable" with an inductive head. All the coding and preequalization strategies used with inductive heads are equally applicable to shielded MRHs. Concomitantly, all the postequalization and digital bit detection strategies used with inductive heads are appropriate to shielded MRHs. Clearly, this equivalence has greatly facilitated the introduction of shielded MRHs.

Shielded MRHs are the most widely used MRH design in computer peripheral recorders. They were first used in 0.5-in. IBM tape drives in 1982, IBM small-diameter disk drives in 1991, and 0.25-in. (QIC) tape drives in 1994. In the 0.5-in. drives, the shields are made of NiZn ferrite, which is inherently resistant to mechanical wear. In the small disk and QIC heads, both permalloy and AlFeSil (Sendust) shields are used. It is usual to place the hard, wear-resistant AlFeSil shield on the upstream edge of the MRH. This minimizes the likelihood of the occurrence of gap-smearing damage.

Shielded MRH Designs

The first shielded MRH designs proposed had the MRE, and its attendant biasing structures, in the gap of a writing head. Thus, the idea was that poles of the writing head, P1 and P2, should double as the shields S1 and S2.

Unfortunately, with this early design, it proved to be almost impossible to maintain the initialization of the MRE. The relatively high deep gap fields, H_0, used in writing heads disturb the desired MRE magnetization state. Deep gap fields are, typically, about equal to three times the coercivity of the recording medium.

The next idea was to fabricate two separate structures. With the writing head poles P1 and P2 entirely separated from the MRH shields S1 and S2, excellent magnetic isolation between the high writing fields and

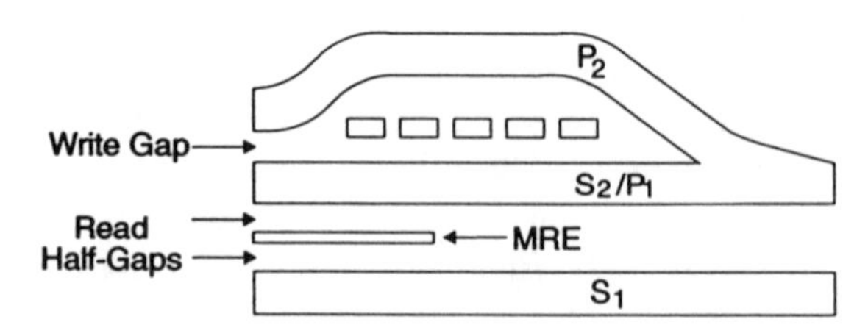

Fig. 8.11. A cross section of a thin-film, merged inductive write head and shielded MRH structure.

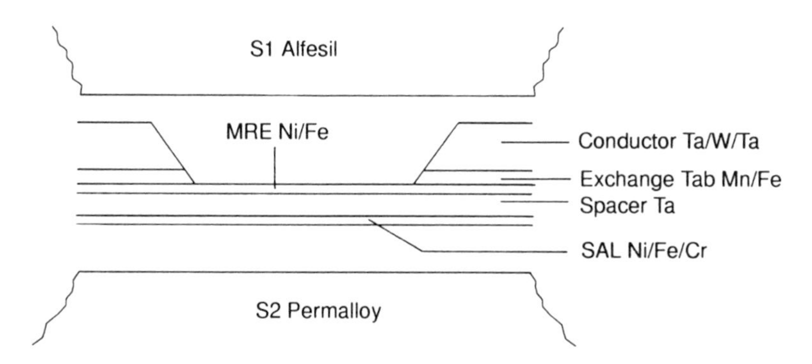

Fig. 8.12. The detailed structure of a SAL-biased, shielded MRH that uses exchange tabs.

the MRE could be achieved. These structures were referred to as *piggyback* heads.

Later, it was realized that a partial separation of the writing and MRH head was sufficient and the two heads could share a common pole. In these heads, one of the write-head poles doubles as one of the shields. This technique, shown in Fig. 8.11, is called *merging* and the resultant heads are referred to as *merged heads*.

In practice, it is always desirable to deposit the MRE thin films as soon as possible while the structure is still almost flat or planar. Thus, the usual order of deposition of the poles is S1, S2/P1, and P2 as Fig. 8.11 indicates.

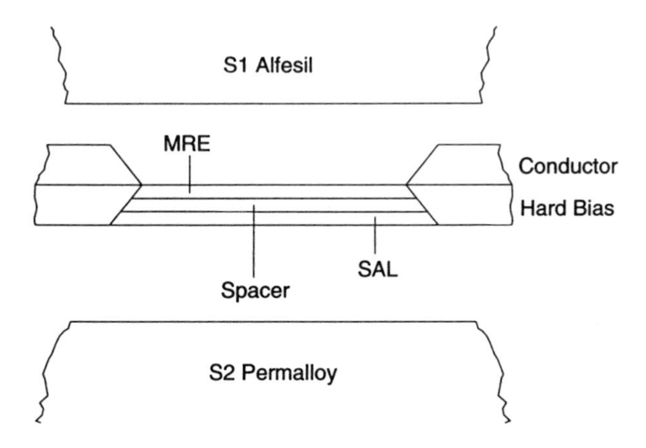

Fig. 8.13. The detailed structure of a SAL-biased, shielded MRH that uses hard bias.

In Fig. 8.12, the overall arrangement of the various parts of an IBM "Corsair"-type disk file shielded MRH is shown. Note the use of exchange tabs for horizontal bias and the use of SAL for vertical bias. The SAL layer is made of Ni/Fe/Cr, an alloy with a higher resistivity than permalloy. The insulator layer is made of a high-resistance phase tantalum. The conductors are composed of low-resistance phase tantalum and tungsten layers.

Figure 8.13 shows the configuration of later generation IBM disk file heads such as "Allicat," "Spitfire," and "Starfire." Note that the exchange tabs have been replaced with permanent magnetic hard bias films abutting both the MRE and SAL. In this way, the edge domains of both the MRE and the SAL are controlled.

Further Reading

Additional reading material is listed here that will prove helpful to readers who seek more detailed information.

Bajorek, C. H., Krongelb, S., Romankiw, L. T., and Thompson, D. A. (1974), "An Integrated Magnetoresistive Read, Inductive Write High Density Recording Head," in *20th Annual AIP Conf. Proc.* American Institute of Physics, p. 24.

Potter, Robert I. (1974), "Digital Magnetic Recording Theory," *IEEE Trans.* **MAG-10,** 3.

CHAPTER

9

Flux-Guide and Yoke-Type Magneto-Resistive Heads

Two persistent drawbacks encountered with single-element shielded magneto-resistive head (MRH) designs are their sensitivity to frictional heating effects and the problem of electrical shorting to conductive recording media. When the tape or disk rubs across the top of the head, the frictional force times the rubbing distance is equal to the work performed and heat generated. The thermal coefficient of resistance of permalloy (0.3%/°C) is such that a 10°C rise in temperature produces a change in resistance equal to the maximum magneto-resistive change ΔR. This phenomenon, termed *thermal spikes,* is well understood. Typically, the heating cycle is very short (a few nanoseconds) and the cooling cycle, controlled by thermal diffusion, is rather long (a few tens of microseconds). The effect of thermal spikes can be eliminated completely by using some of the double magneto-resistive element (MRE) designs discussed next in Chapter 10. It can be minimized by careful design of the spectral content of the channel codes and the postequalization strategies used.

Flux-guide and yoke-type MRHs offer yet another way to eliminate completely thermal spike effects. This is accomplished by moving the thermally sensitive MRE away from the head–medium contact area at the top of the head.

The problem of the electrical shorting of the MRE measuring current to electrically conductive recording media, such as thin-film rigid disks, has to be addressed by the system designer. It can be ameliorated, for example, by "floating" the disks. Floating means electrically isolating the disks and maintaining them at the same electrical potential as the MREs. The MRE of flux-guide and yoke-type MRHs has to be electrically insulated from the yoke structure in order to prevent the measuring current from spreading into the yoke. This electrical insulation solves the head–disk shorting problem.

Flux-Guide MRHs

A flux-guide design is shown in Fig. 9.1. The MRE spans or bridges a gap in a highly permeable flux guide that is placed between the highly permeable shields S1 and S2.

The mode of operation is self-evident. The MRE senses the flux flowing down the center flux guide just as does the MRE in a conventional design of a shielded MRH. It follows that the output voltage spectral shape and pulse shape of a flux guide MRH are the same as those of a shielded MRH.

A major disadvantage of this design, and all the other flux-guide and yoke-type MRHs discussed in this chapter, is that the flux efficiency cannot be high. By its very nature, an MRE is a thin film of high magnetic reluctance. Moreover, it is necessary to provide electrically insulating gaps between the MRE and the flux guide.

In the last chapter, the conventional shielded MRH was discussed as a magnetic transmission line. The flux-guide MRH's flux-guide/shield system is likewise a flux transmission line. To achieve a high efficiency in a flux-guide MRH, the flux guide must be thick and shallow, so that its reluctance is negligible. On the other hand, it is necessary to keep the shield-to-shield distance ($2g + T$) small in order to have good spatial

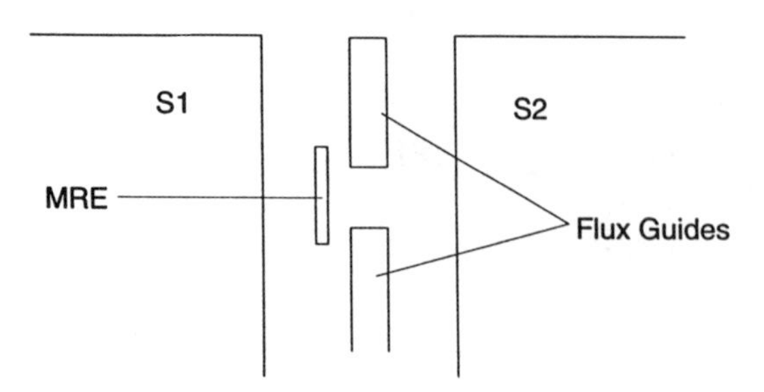

Fig. 9.1. The structure of a flux-guide shielded MRH.

resolution (small PW_{50}) so, unfortunately, the flux guide cannot, therefore, be thick. Moreover, the depth of the flux guide has to be large enough so that the thermal impulses generated on the top surface have diffused sufficiently before reaching the MRE. The net result is that it is difficult to achieve flux efficiencies higher than 20%. Accordingly, this flux-guide design produces much lower specific output voltages than does a conventional shielded MRH.

Yoke-Type MRHs

The basic yoke-type MRH design is shown in Fig. 9.2. A deposited thin-film yoke structure is shown with the lower yoke Y1 separated into two parts and bridged by the MRE. Again, by having the MRE some distance from the recording medium contact area and by having the MRE electrically insulated from the yoke, the two problems of thermal spikes and electrical shorting are resolved.

Once more, because the MRE is thin and has electrically insulating gaps, the magnetic reluctance of the structure is high and the flux efficiency of the head is low. Even with careful design, it is difficult to achieve even 20% efficiency in yoke-type MRHs.

The yoke-type MRH senses the ring head flux directly and its output voltage spectrum is that given in Eqs. (3.3) and (8.5). Where shielded MRH or inductive head produces the Lorentzian pulse shape $L(x)$ and is proportional to $d\,M(x)/dx$, the yoke-type design generates an output which is given by the integral of $L(x)$ and is proportional to $M(x)$. In some recording systems, the inductive head output is integrated and the yoke-type MRH output suits such systems directly.

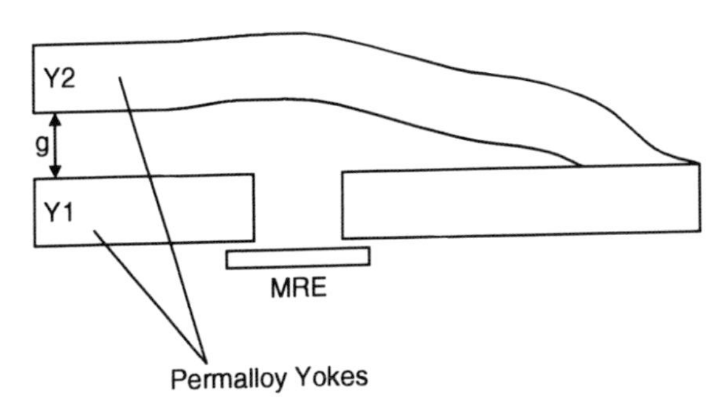

Fig. 9.2. The structure of a yoke-type MRH.

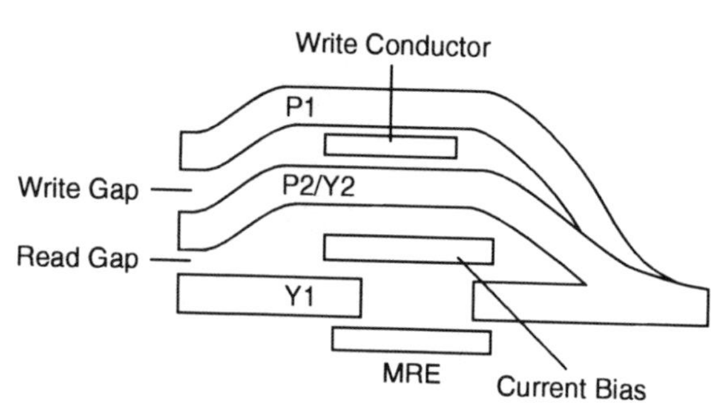

Fig. 9.3. The structure of a thin-film, merged, inductive writing head and yoke-type MRH.

The complete merged thin-film write head and yoke-type MRH structure used in the Philips DCC is sketched in Fig. 9.3. Again the MRE is the first layer to be deposited and the order of deposition of the permalloy yokes and poles is Y1, Y2/P2, and P1. Note the placement of the single-turn, servo-bias current conductor and the use of a single-turn write conductor. When fabricating a 16-channel write–read head structure for 4-mm-wide tape, every part of the structure should be as simple as is possible. A similar philosophy guides the 18- and 36-channel IBM 0.5-in. tape where the merged write heads have two turns only.

Further Reading

Additional reading material is listed here that will prove helpful to readers who seek more detailed information.

Zieren, V., Somera, G., Ruigrok, M., de Jongh, M., van Straaven, A., and Folkerts, W. (1993), "Design and Fabrication of Thin Film Heads for the Digital Compact Cassette Audio System," *IEEE Trans.* **COMM-29,** 6.

CHAPTER

10

Double-Element Magneto-Resistive Heads

Many of the disadvantages inherent in the single-element shielded magneto-resistive head (MRH) can be alleviated or avoided completely by using MRH structures with two active magneto-resistive elements (MREs). Among these disadvantages are the susceptibility to thermal spikes, the pulse amplitude asymmetry, the side-reading asymmetry, and the electrical shorting to disk problem.

In double-element MRHs, each individual MRE operates as described earlier in this book. The advantages of double-element MRHs are associated with the details of the electrical connections made to, and the directions of magnetization of, the MREs.

Figure 10.1 shows two types of double-element MRHs. In Fig. 10.1(A) the two MREs are unshielded, whereas in Fig. 10.1(B) the two MREs are between two shields S1 and S2. It has already been pointed out that the spectral phase of unshielded vertical MRHs and shielded MRHs is the same. Thus, both Eqs. (7.5) and (8.8) contain the factor j $(=\sqrt{-1})$, which denotes 90° phase shift from the recorded magnetization. It follows that each equation shows the same "digital" output pulse symmetry shown in Figs. 3.4 and 3.5. This pulse basically follows the differential of the magnetization transition, $dM(x)/dx$ and has the Lorentzian form of $L(x)$.

When two MREs are used, there are only two possible output pulse shapes. When two MRE output pulses of the same polarity are added the

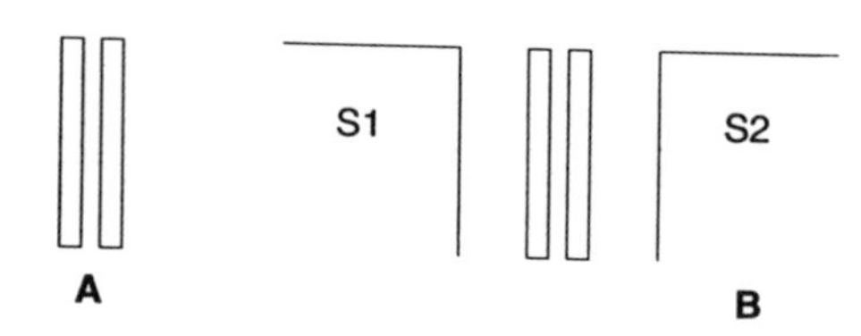

Fig. 10.1. (A) A double-element unshielded MRH and (B) a double-element shielded MRH.

result has the same dM/dx shape. On the other hand, when two MRE pulses of the same polarity are subtracted, the result has a different shape. Recalling the signal processing engineer's method of performing differentiation discussed in Chapter 8, the different shape is bipolar, following d^2M/dx^2 or dL/dx.

Of course, when there are two MREs in proximity to each other, mutual magnetic interaction effects occur and the actual pulses produced are not exactly those given by simple addition and subtraction of the single MRE pulses. The details of the mutual interactions are beyond the scope of this book. Whatever the mutual interactions may be, however, to first order the two possible output pulses of a pair of MREs are dM/dx and d^2M/dx^2.

Classification of Double-Element MRHs

A pair of MREs is shown in Fig. 10.2. Note that in this pair, the magnetization vectors M are parallel and the current flows are parallel. The magnetization vector directions have been established by some verti-

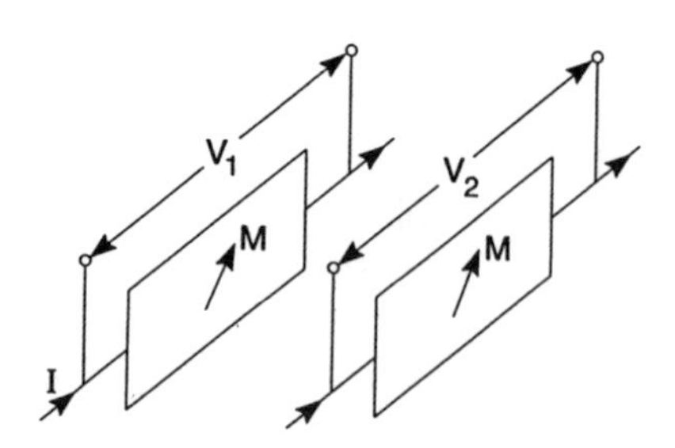

Fig. 10.2. Schematic of two MREs showing their current and magnetization directions and the MR voltages developed.

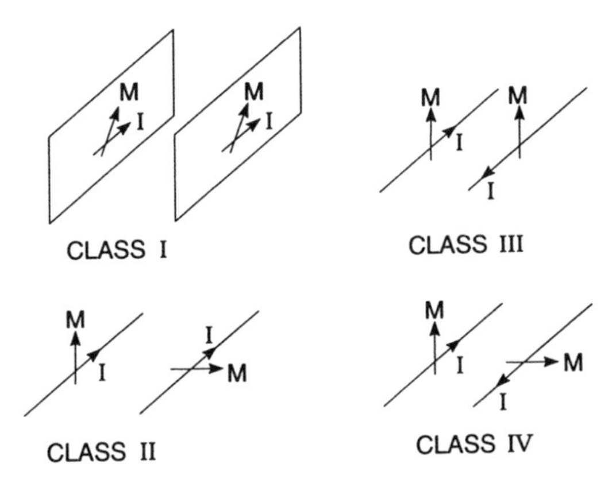

Fig. 10.3. Four classes of double-element MRHs, based on the possible symmetries of the current and magnetization directions.

cal bias technique, which is not shown. The current directions are established by means of external conductors, which are also not shown. Finally, the variable parts of the MRE voltage, V_1 and V_2 are presumed to be sensed either in addition, $V_1 + V_2$, or subtraction, $V_1 - V_2$, by a suitable electrical connection which, again, is not shown. Subtractive sensing is often called "differential" sensing for obvious reasons.

With two possible current directions, two possible magnetization directions, and two voltage sensing modes, there are eight combinations to consider. The four magnetization and current direction possibilities are

TABLE 10.1

Class	Electrical sensing: Addition	Electrical sensing: Subtraction
I	$V = \frac{dM}{dx}$	$\frac{dV}{dx} = \frac{d^2M}{dx^2}$
II	$\frac{dV}{dx} = \frac{d^2M}{dx^2}$	$V = \frac{dM}{dx}$
III	$V = \frac{dM}{dx}$	$\frac{dV}{dx} = \frac{d^2M}{dx^2}$
IV	$\frac{dV}{dx} = \frac{d^2M}{dx^2}$	$V = \frac{dM}{dx}$

shown in Fig. 10.3 and they are classified here, arbitrarily, as Classes I, II, III, and IV. The eight possible combinations are presented in Table 10.1.

Note that the Class I addition configuration yields the usual unipolar output pulse, dM/dx or the Lorentzian $L(x)$. Class I subtraction is called a "gradiometer" and produces d^2M/dx^2 or dL/dx bipolar pulses.

The Class II addition connection scheme is called, by Kodak, the "Dual Magneto-Resistor" (DMR) head and it has the bipolar d^2M/dx^2 type output pulse. The Class II subtraction arrangement gives the normal unipolar dM/dx pulse and is called, by Hewlett-Packard, the "Dual-Stripe" head.

It is of interest to note that the remaining four entries in Table 10.1 have not yet received names.

Advantages of Double-Element MRHs

All of the configurations of double-element MRHs which use subtractive or differential sensing are inherently robust against thermal spikes. This is because the differential sensing rejects all common-mode resistance changes such as those caused by (the almost) synchronous thermal fluctuations in the MREs.

The configurations which eliminate the fundamental even harmonic distortion or pulse amplitude asymmetry are Class II and Class IV. To achieve pulse amplitude symmetry, it is necessary to have differing magnetization directions.

Both of the Classes in which the magnetizations are not parallel exhibit, to first order, symmetrical side-reading or off-track responses. Should absolutely symmetrical behavior be required, it can be achieved, in principle, by changing the magnetization angles from the pattern shown in Fig. 10.4(A) to the configuration shown in Fig. 10.4(B). Perfect off-track symmetry is expected in all of the four possible patterns which have their z and y magnetization components anti-parallel and parallel respectively, as occurs in Fig 10.4(B).

The configurations in which the current in the MREs flows in opposite directions (Classes III and IV) are attractive. They offer an easy way to alleviate, but not eliminate, the electrical shorting to disk problem. When one end of each MREs is connected and held at zero, or ground, potential, the average potential of the complete double-element structure is also zero. This, of course, means the device requires three external

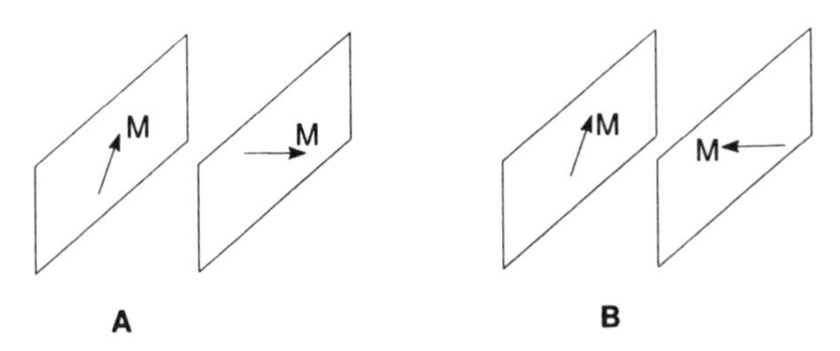

Fig. 10.4. (A) Up–down magnetization giving low side-reading asymmetry and (B) Up–down magnetization giving zero side-reading asymmetry.

connections or terminals. The same zero average potential result can, of course, be achieved when the currents flow parallel, but in this case four external connections are needed.

General Comments

Double-element MRHs offer many advantages over single-element MRHs. These advantages are achieved at the expense of considerably increased device and circuit complexity. It follows then that the system designer has to make choices. These choices can only be made after obtaining knowledge of the relative importance of the problems in particular recording systems. For example, electrical shorting and side-reading symmetry might be of little importance in a relatively wide track-width tape recorder using a magnetic coating of low electrical conductivity. Thermal spikes might be considered of little importance in flying-head disk drives because of their relative infrequency.

Adding to the difficulty of selecting the best design are, naturally, the complicated issues of device fabrication and process yields, both of which are beyond the scope of this book.

Nevertheless, some general comments are possible. For example, the question of whether the unipolar Lorentzian pulse $L(x)$ is to be preferred over its bipolar Lorentzian $dL(x)/dx$ depends on the origin of the major noise sources in the recording system. If the major noise source is the recording medium itself, then there is little to choose between the two possibilities because either pulse can be transformed, by extremely simple external circuitry, into the other. Moreover, the transformations (differentiation and integration) are linear and reversible so that no information is, in principle, lost in the process.

When, however, the major noise source occurs later in the reading cascade, for example, in the reading head or first-stage read amplifier, the ability to perform signal transformations by means of the spatial design of the reading head before the noise can be extremely useful. An analogous situation arises in the yoke-type (DCC) MRHs discussed in Chapter 9, where the output is proportional to $M(x)$ or the integral of $L(x)$.

Some of the configurations shown in Fig. 10.3 appear to be easier to implement than others. For example, in the Class II and III configurations, the magnetization directions can be established by using MRE currents alone. Other configurations require the provision of vertical bias by one of the other techniques outlined in Chapter 5.

Finally, note that all of the configurations that yield the d^2M/dx^2 output can, in principle, be operated satisfactorily without shields. Recall that the fundamental shunting problem of Hunt's vertical head causes the high-frequency output spectrum to roll off as $1/k$. A second spatial differentiation, which achieves the change from a dM/dx to a d^2M/dx^2 response, is the equivalent of multiplying the spectrum by the factor jk. Obviously, the short wavelength spectrum roll-off is corrected. The remaining factor, j (a 90° phase shift) can then be corrected with an external filter, called an *all-pass phase shifter* or *Hilbert transformer*.

The study of double-element MRHs is a very active area, rife with invention. Time alone will reveal whether or not there is a preferred configuration for specific applications.

Further Reading

Additional reading material is listed here that will prove helpful to readers who seek more detailed information.

Indeck, R. S., Judy, J. M., and Iwasaki, S. (1988), "A Magnetoresistive Gradiometer," *IEEE Trans.* **MAG-24,** 6.

Mallinson, J. C. (1990), "'Gradiometer' Head Pulse Shapes," *IEEE Trans.* **MAG-26,** 2.

Smith, N., Freeman, J., Koeppe, P., and Carr, T. (1992), "Dual Magnetoresistive Heads for High Density Recording," *IEEE Trans.* **MAG-28,** 5.

CHAPTER

11

Comparison of Shielded Magneto-Resistive Head and Inductive Head Outputs

It is of great interest to compare the output voltages produced by shielded magneto-resistive heads (MRHs) and inductive heads because the decision to use a shielded MRH is strongly dependent on this comparison. There are two ways to make this comparison. In the first, the ratio of the small-signal output voltage spectra of a shielded MRH and an inductive head is considered. This ratio is principally of interest in linear, analog recording.

The second comparison considers the case of a shielded MRH in which the MRE element thickness T and depth D have been chosen so that the specific peak voltage of the isolated transition output pulse in digital recording is optimized. The details of this design process are discussed in Chapter 14.

Small-Signal Voltage Spectral Ratio

The small-signal output voltage spectrum of the single-element shielded MRH is, following Eq. (8.8),

$$V_{\mathrm{MRH}}(k) = \frac{1}{2}\left(\frac{I\Delta R}{2}\right)\left[\frac{4\pi M_R(1 - e^{-k\delta})\, e^{-kd}}{k}\right]\left(\frac{I}{H_{\mathrm{bias}}T}\right) \times \left(\frac{\sin kg/2}{kg/2}\right) 2j \sin\left(\frac{k(g + T)}{2}\right). \tag{11.1}$$

The output voltage spectrum of an N-turn inductive head operated at a head-to-medium relative velocity V is from Eq. 3.5.

$$V_{\mathrm{IND}}(k) = j10^{-8}\, NVW\, [4\pi M_R(1 - e^{-k\delta})\, e^{-kd}] \left(\frac{\sin kg/2}{kg/2}\right). \quad (11.2)$$

Cancelling the common terms, the ratio of the shielded MRH to inductive head output voltage is

$$\frac{V_{\mathrm{MRH}}}{V_{\mathrm{IND}}} = \frac{\frac{1}{2}\cdot\frac{I\Delta R}{2}\cdot\frac{1}{k}\cdot\frac{1}{H_{\mathrm{bias}}T}\cdot\frac{2\sin k(g+T)}{2}}{10^{-8}\, NVW}. \quad (11.3)$$

This ratio can be evaluated by applying the following approximations. The measuring current is usually limited by the electromigration phenomenon discussed in Chapter 14 and is J_0TD where the current density, J_0, can be taken to be 10^7 A/cm^2. The maximum change in the resistance of the magneto-resistive element (MRE), ΔR, is equal to $\Delta\rho W/TD$, where, in permalloy, $\Delta\rho = 4\%$ of the resistivity, 20×10^{-6} Ω-cm. In many designs, the depth D of the MRE is about ten times its thickness T, so that $T/D = 0.1$. The vertical bias field, H_{bias}, must be approximately equal (and opposite to) the average vertical demagnetizing field in the MRE, $4\pi M_S \sin\theta\cdot T/D$. In permalloy, $4\pi M_S$ is approximately 12,500 G. The biased magnetization angle θ is close to 45°. Finally, we assume that $g \approx 5T$ and that the products kg and $k(g + T)/2$ are small.

After making all of the preceding substitutions, Eq. (11.3) becomes simply,

$$\frac{V_{\mathrm{MRH}}}{V_{\mathrm{IND}}} \approx \frac{10^{+6}}{NV}. \quad (11.4)$$

This equation can be restated in a particularly direct form:

$$V_{\mathrm{MRH}} \equiv V_{\mathrm{IND}}\ (NV \approx 10^6 \text{ cm/sec}). \quad (11.5)$$

The small-signal output voltage of a single-element permalloy anisotropic MR shielded MRH is equal to that of an inductive head with an NV product equal to about 10^6 cm/sec.

Consider a typical analog audio recorder where the tape speed, V, is about 10 cm/sec. The shielded MRH small-signal sensitivity is equivalent to that of an $N = 10^5$ turn inductive head. At a tape speed of 100 cm/sec, equivalence occurs with $N = 10^4$ turn inductive heads. Clearly, MRHs are extremely sensitive transducers at long wavelengths.

Large-Signal Voltage Comparison

In digital recorders, for example, hard-disk drives, the remanence of the recording medium is usually saturated. It follows that the fields and fluxes emanating from the magnetization transition are not small signals.

In these cases, it is necessary to design the shielded MRH carefully so that the quasi-linear range of operation of the MRE about the vertical bias point is not exceeded. In terms of magnetization angle changes, typically excursions of $\pm 30°$ about the quiescent vertically biased 45° are acceptable. Beyond 30°, the second harmonic generation or pulse asymmetry incurred is usually deemed to be unacceptable even for digital recording. As discussed in Chapter 14, the flux capacity of the MRE must match the remanence times thickness product of the recording medium, thus $B_s T \approx 2M_R\delta$.

In an optimally designed MRH using permalloy, the peak pulse voltage per unit track width is a constant, as has been mentioned previously in Chapter 8.

Consider a shielded MRH where the MRE's $D < l$, the transmission line characteristic length. The flux efficiency of this MRH is about 1/2. Of the total magneto-strictive dynamic range of 4%, only $\pm 2\%$ is available for peak positive and peak negative going signals, respectively. If the current density is J_0, the measuring current, I, is $J_0 TD$ and the output voltage $\delta V = \frac{1}{2} I \delta R$ where δR is 2% of $R = \rho W/TD$. Note that the MRE's depth D and thickness T do not influence the voltage δV in this optimized design case. The output voltage is, for a current density $J_0 = 10^7$ A cm^{-2},

$$_{\text{opt}}V_{\text{MRH}} = 0.5 \times 10^7 (2/100)(20 \times 10^{-6}) W = 2W \quad \text{[volts]}. \qquad (11.6)$$

When considering shielded MRHs that have been optimally designed, it is not possible to make universal comparisons with inductive heads. Whereas the optimized MRH has a constant specific output voltage, the voltage produced by a linear inductive head is always proportional to the $M_R\delta$ product of the recording medium.

For thin-film disk and metal evaporated tape recording media, however, the $M_R\delta$ products are comparable. In a typical thin-film disk, the M_R is about 800 emu/cm^3 and the thickness δ is about 2 μin. (500 Å). A modern ME tape coating may be taken to have $M_R = 200$ emu/cm^3 and $\delta = 8$ μin. (2000 Å). For such media, the ratio of the optimized shielded MRH to inductive head peak pulse output voltage is shown in Chapter 14 to be

$$\frac{{}_{\text{opt}}V_{\text{MRH}}}{V_{\text{IND}}} = \frac{80{,}000}{NV}. \tag{11.7}$$

In this specific case then, the optimized shielded MRH output is equal to that of an inductive head with an NV product of 80,000 cm/sec. This equivalence is shown in Table 11.1. This table shows several combinations of inductive head numbers of turns N and head-to-medium velocities V whose product equals 80,000 cm/sec.

Many useful observations can be made from Table 11.1. First, we can see that an MRH is of little interest at the highest speeds ($V = 4000$ cm/sec) because it is relatively easy to construct inductive heads with 20 turns. Such speeds occur in professional videotape recorders (VTRs). Second, it is equally obvious that MRHs are of great interest at the lowest speeds because it is difficult to make inductive heads that carry many thousands of turns.

The first commercial application of MRHs in the 1970s occurred in bank and credit card readers, where the head-to-medium velocity is, typically, only a few centimeters per second. The first application of MRHs in computer peripherals in the 1980s was in the IBM 3480 0.5-in. digital tape drive where the tape speed is, typically, approximately 250 cm/sec. It is more feasible to make an 18-track head structure with shielded MRHs than with 320-turn inductive heads.

TABLE 11.1

Digital recorders	V cm/sec × turns
Digital VTRs	4000 cm/sec × 20T
5.25-in. disk (inner track)	2000 cm/sec × 40T
2.5-in. disk (inner track)	1000 cm/sec × 80T
0.5-in. digital tape	250 cm/sec × 320T
Digital audio tape	25 cm/sec × 3200T
ATM and bank card readers	2.5 cm/sec × 32,000T

The middle region of Table 11.1 is of great interest because it encompasses the range of circumferential velocities found in the small-diameter disk drives used in today's portable computers. The circumferential velocity is $V = r\omega$ where r is the track radius and ω is the angular velocity of the disk. The angular velocity is equal to 2π times the number of revolutions per second.

Suppose the effective limit on the number of turns that can be fabricated economically in a thin-film head is $N = 50$. Then, any proposed head–disk application where the track velocity is less than 1600 cm/sec is a potential candidate for an optimized shielded MRH, because it will produce a higher output voltage.

Taking the rotational speed of small disk drives to be 100 revolutions per second (6000 revolutions per minute), we see that whereas there can be little interest in using a shielded MRH for a 5.25-in.-diameter disk drive, there is little doubt that it is the preferred choice for 2.5-in.-diameter and smaller disk drives.

Note carefully that the $NV = 80{,}000$ cm/sec comparison we arrived at earlier is based on recording media of a particular $M_R\delta$. The thinner the medium, the more attractive becomes the optimized MRH when compared to an inductive head.

A thinner medium mandates that the optimized design have a thinner (smaller T) and shallower (smaller D) MRE. As shown in Chapter 14, the thickness T has to be directly proportional to the medium's thickness δ. The MRE depth D must change as $\sqrt{\delta}$ because $l \propto \sqrt{T}$.

As the MRE becomes thinner, the magneto-resistive coefficient, $\Delta\rho/\rho_0$, decreases because surface scattering of the conduction electrons, which increases ρ_0, becomes proportionately more important. With $T =$ 250 Å, $\Delta\rho/\rho_0$ is only 2% in permalloy. A practical limit for permalloy sensors is about $T \approx 150$ Å, corresponding to an 800 emu-cm^{-3} medium with $\delta \approx 100$ Å.

Further Reading

Additional reading material is listed here that will prove helpful to readers who seek more detailed information.

Potter, Robert I. (1974), "Digital Magnetic Recording Theory," *IEEE Trans.* **MAG-10,** 3.

CHAPTER

12

The Giant Magneto-Resistive Effect

The giant magneto-resistive (GMR) effect was unknown until 1988. It is an entirely different phenomenon from the anisotropic magneto-resistive (AMR) effect treated previously in this book.

GMR was discovered because developments in high vacuum and deposition technology made possible the construction of molecular beam epitaxy (MBE) machines capable of laying down multiple thin layers only a few atoms thick. Such structures are often called superlattices because the periodicity mimics, on a large scale, crystal lattices.

Superlattices

The availability of MBE machines led to the investigation of the exchange coupling between a multiplicity of thin layers of a ferromagnet (Fe) separated by thin layers of a "nonmagnetic" material (Cr). Chromium is, of course, not actually nonmagnetic, but is an antiferromagnet when below its Néel temperature of 37°C. Typically, the layer thicknesses used were a few tens of angstroms as shown in Fig. 12.1.

The specific aspect of the interlayer exchange studied was the change from ferro-magnetic (parallel) to antiferromagnetic (antiparallel) coupling as the chromium layer thickness is varied. The oscillatory variation found is shown in Fig. 12.2.

Measurements of the electrical resistance showed that in cases where

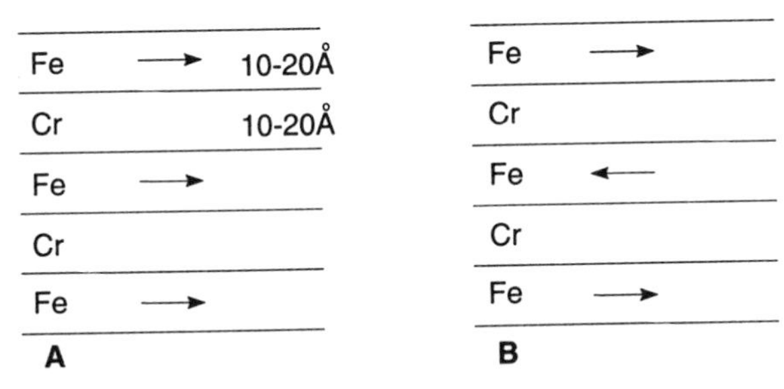

Fig. 12.1. An iron-chromium superlattice with (A) ferromagnetic (parallel) and (B) antiferromagnetic (antiparallel) exchange coupling.

the exchange was antiferromagnetic, the resistance could be changed as shown in Fig. 12.3, by applying large (10,000 Oe) magnetic fields. Because the resistance change was very large, $\Delta\rho/\rho_0 \approx 40\%$, this effect was called the *giant* magneto-resistive effect.

Studies showed that the electrical resistance was high whenever the iron layer magnetizations were antiparallel and low when they were forced, by the external field, to be parallel. The variation in resistance was found experimentally to be proportional to $-\cos\beta$, where β is the angle between the adjacent layer magnetizations. Note carefully that the GMR is independent of the measuring current direction and that this is in contrast to the AMR effect, where the magnetization–current angle θ is the important factor.

Despite the excitement the discovery of GMR caused among material scientists and physicists, those involved in MRH design did not consider the effect useful for two reasons. Not only were extremely high external fields (10,000 Oe) required to change the resistance, but also, as shown in

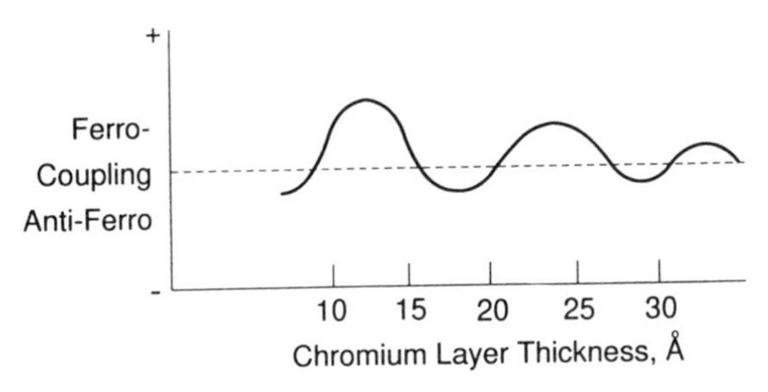

Fig. 12.2. The periodic variation of the exchange coupling in an iron-chromium superlattice versus the chromium layer thickness.

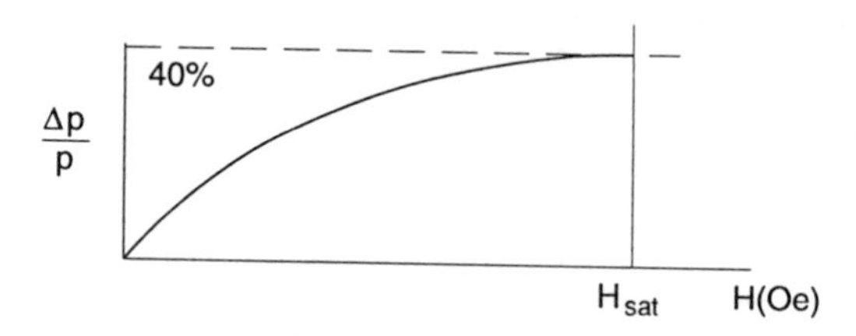

Fig. 12.3. The GMR coefficient of an antiferromagnetically coupled Fe-Cr superlattice versus field.

Fig. 12.4, the low field sensitivity $[d(\Delta\rho/\rho_0)/dH]$ of AMR in permalloy was much higher than that displayed in the Fe-Cr superlattice GMR.

Researchers quickly realized that these shortcomings were attributable to two facts: first, the high anisotropy field ($2K/M \approx 560$ Oe) of the iron layer and, second, the use of chromium, an antiferromagnet, in the interlayers, which causes strong exchange coupling. A high external field is needed because it has to overcome both the iron layer anisotropy field and the antiferromagnetic exchange field in order to force the iron layer magnetizations to be parallel.

The solution is to use permalloy layers separated by a truly nonmagnetic metal, for example, copper, silver, or gold, as shown in Fig. 12.5. Permalloy has, as has been discussed in Chapter 4, an extremely low anisotropy field ($\approx$5–10 Oe) at room temperature. Moreover, when Cu, Ag, or Au interlayers are used there is only low, or in some cases zero, exchange coupling between the permalloy layers. The precise interfacial conditions, which govern the magnitude of the exchange coupling field, are not known. Most of the reasons, outlined in Chapter 5 in the discus-

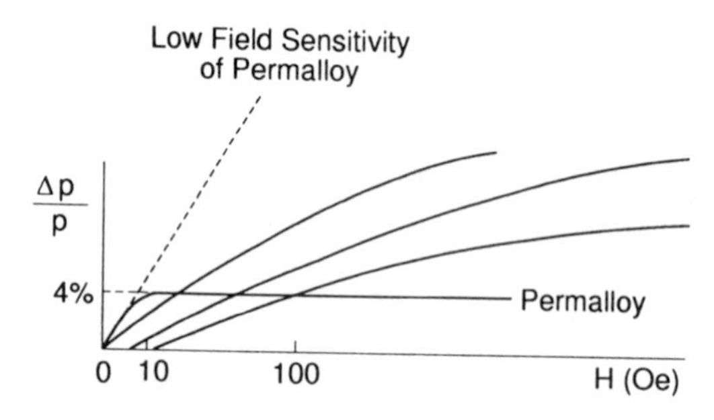

Fig. 12.4. The low-field GMR coefficient of an Fe-Cr superlattice compared to AMR in permalloy.

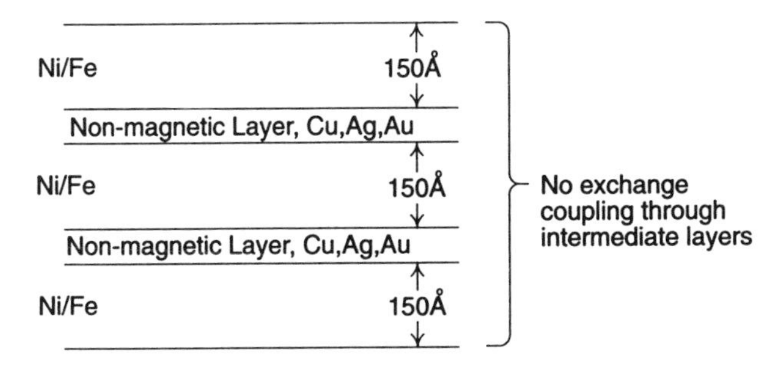

Fig. 12.5. A permalloy spin valve structure with little or no exchange coupling through the nonmagnetic intermediate layers.

sion of exchange bias, for the imprecise knowledge of the magnitude of exchange anisotropy apply equally in the superlattice coupling problem. Again, because the exchange coupling between atoms is critically dependent on their spacing, it may well transpire that it is not possible to measure the interfacial structure in sufficient detail to predict accurately the exchange coupling achieved.

Because the external field now has only to be greater than the low anisotropy field and the low exchange field, the permalloy layers can be switched from the antiparallel, high-resistance state to the parallel, low-resistance state by a relatively small field. The resistance change characteristic shown in Fig. 12.6 is typical. Note that now the GMR low-field sensitivity is now larger than that found in AMR permalloy.

The Physics of the GMR Effect

Despite the novelty of GMR, simple explanations of the effect have been forthcoming. The phenomenon is believed to depend principally on

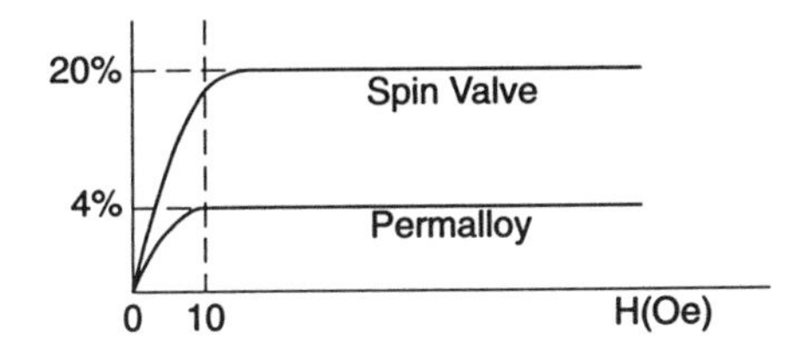

Fig. 12.6. The GMR of a permalloy spin valve compared with the AMR of permalloy.

the selective scattering of conduction electrons at the interfaces between the ferromagnetic and nonmagnetic layers.

The conduction electrons in iron and permalloy are the $3d$ electrons discussed in Chapter 1. Every electron has the quantum-mechanical property called spin. To minimize energy, the majority of the electron spin directions in the ferromagnet are oriented parallel to the magnetization vector M.

When an electric field is applied, the spin-oriented conduction electrons accelerate until they encounter a scattering center. The existence of scattering centers is, of course, the origin of the electrical resistivity. The average distance the conduction electron accelerates is called the *coherence length* and it is of the order of a hundred angstroms in the metals used in GMR superlattices.

Consider the case in which the scattered electron traverses the interlayer of a nonmagnetic metal. Provided the interlayer thickness is less than the coherence length, the conduction electron arrives at the interface of the adjacent ferromagnetic layer still carrying its original spin orientation. This process is shown in Fig. 12.7.

When the adjacent magnetic layers are magnetized in a parallel manner, as in Fig. 12.7(A), the arriving conduction electron has a high probability of entering the adjacent layer with negligible scattering, because its spin orientation matches that of that layer's majority spins. On the other hand, as shown in Fig. 12.7(B), when the adjacent layer is magnetized in an antiparallel manner, the majority of the spin-oriented electrons suffer strong scattering at the interface because they do not match the majority spin orientation. When the magnetic layers are in the ferro state (magne-

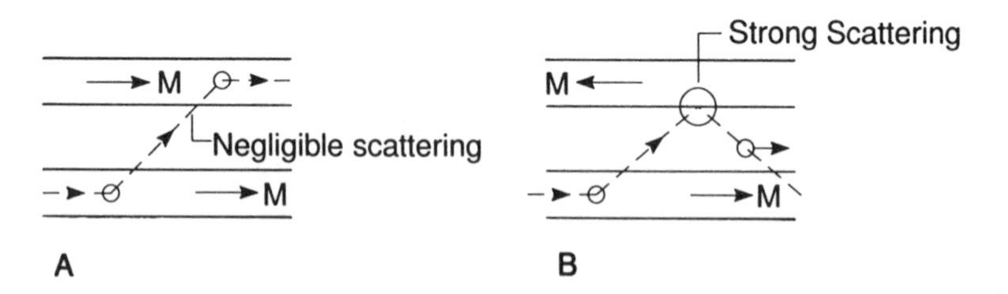

Fig. 12.7. The scattering of a spin-polarized, conduction electron at the interface of (A) a ferromagnetically and (B) an antiferromagnetically coupled layer.

tized parallel) the resistance is low and vice versa in the antiferro state (magnetized antiparallel).

The precise details of the interface which controls the magnitude of the strong scattering are not known in detail. It is known that the surface roughness of the interface is critical. As measured by low-angle X-ray diffraction, roughnesses of a few Å give the greatest scattering and, thus, highest GMR. Bearing in mind that atom-to-atom spacings are in the range of 2 to 3 Å in these materials and the fact that X-ray diffraction measurements are averages of the roughness over the X-ray beam projected area (typically, many square micrometers), we may well wonder just precisely what a roughness of 2 to 3 Å really means. Once more, as in exchange coupling at interfaces, the atomic details necessary for the development of high GMR may be beyond experimental determination.

Despite the uncertainty in the exact interfacial structure required, it is possible to list several necessary conditions for the development of GMR.

First, the two materials used in the superlattice must be immiscible. In other words, they must not mix or dissolve in each other, just as oil and water are not soluble in each other. If the materials are miscible, the interfaces, and eventually the interlayers, can be expected to diffuse into each other over time and thus become no longer sharply defined.

The next prerequisite for GMR is that the ferromagnetic layers must have some mechanism, whether exchange coupling or mere magnetostatics, that establishes the antiferro (antiparallel) magnetization state in zero external field.

Third, the interlayer of nonmagnetic material must be thinner than the conduction electron coherence length. This is so that the electrons can "remember" their spin orientations as they cross between the ferromagnetic layers. With coherence lengths of a hundred Å, this means that nonmagnetic interlayer thicknesses of a few tens of Å are required.

In each ferromagnetic layer, two spin orientations of the conduction electrons coexist. The majority are parallel to the magnetization and the minority antiparallel. This fact has led to the idea of a "two-current" GMR model.

The two-current model is shown in Fig. 12.8. It is simply a two-leg, four-resistor parallel network. In the low-resistance state shown in Fig. 12.8(A), the opposite legs of the parallel circuit have two low and two high resistances, respectively. The two currents, electron spin-up and

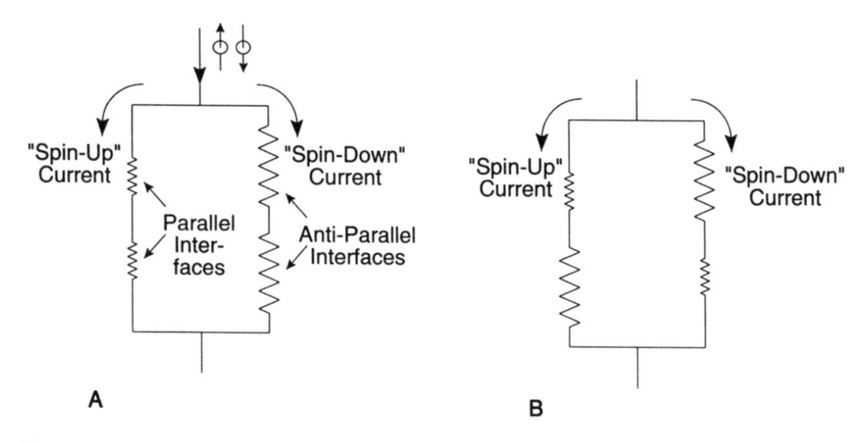

Fig. 12.8. The two-current model electrical circuit showing (A) the low-resistance and (B) the high-resistance conditions.

spin-down, respectively, enter at the top. The spin-up current flows down the left leg, which has two low resistances corresponding to two magnetization parallel interfaces. The spin-down current flows through the two high resistances in the right leg that models two antiparallel interfaces.

In Fig. 12.8(B), the high-resistance state of the two-current model is shown. Now, in whichever leg the two spin-oriented currents flow, they encounter equal resistance. There is no easy, low-resistance leg of the circuit for either the majority or minority electrons.

As is the case in explanations of AMR which invoke the statistics of the electron energies, a band structure or density of states explanation of GMR can be made. Such arguments are frequent in pedagogical expositions. However, they lack real utility because they do not include the structural detail that is known to be necessary for the development of high GMR.

Granular GMR Materials

Shortly after the discovery of GMR in superlattices, analogous discoveries were made in granular, or heterogeneous, GMR materials.

The first such material was made by sputtering alternating, relatively thick layers of cobalt and copper. Such a layered structure of two immiscible metals is in thermodynamic equilibrium because the high-energy interfacial area is minimized. Recall that after vigorous shaking of oil and

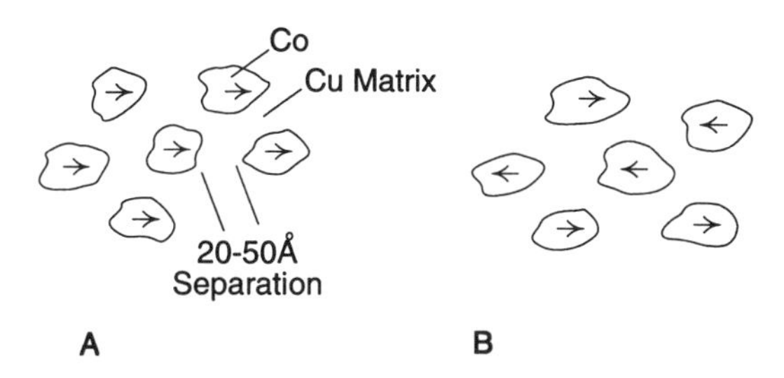

Fig. 12.9. The magnetization state of a granular cobalt–copper two-phase GMR material in (A) the low-resistance and (B) the high-resistance states.

water, the small oil droplets rapidly coalesce into layers in order to reduce the oil/water interfacial area and energy.

The sputtered multilayer of cobalt and copper was then annealed, during which process the layers of cobalt diffuse to become irregular particles or grains in a matrix of copper as shown in Fig. 12.9. An optimum annealing time and temperature exists (about 300°C for 1 hour) that produces the desired microstructure where the cobalt grains are separated by distances less than the coherence length of electrons in copper.

Unfortunately, the low field sensitivity $d(\Delta\rho/\rho_0)dH$ of these first granular GMR materials was much too low for MRH device applications. Again the problem was in the improper selection of the ferromagnetic material, cobalt, which has an extremely large anisotropy field ($2K/M \approx$ 6000 Oe). Moreover, the cobalt grain sizes achieved after the annealing were much too small, being well into the superparamagnetic range (a few tens of Å). For these two reasons, large external fields ($\approx$10,000 Oe) were needed to force the cobalt grains into the parallel magnetization state.

The solution to the low-sensitivity problem is obvious. Replace the cobalt with permalloy and make the grain sizes larger. Accordingly, in the most recent granular GMR materials, the starting point is a sputtered multilayer of permalloy and silver. Unlike copper, silver is not miscible in permalloy. After suitable annealing (again at 300°C for 1 hour) the silver diffuses down the grain boundaries (but not into the grains themselves) resulting in the structure shown in Fig. 12.10.

Now, because the low anisotropy permalloy grains are not superparamagnetic, the required saturation external magnetic field is of the order of 10 Oe only, and a large low-field GMR sensitivity can be realized.

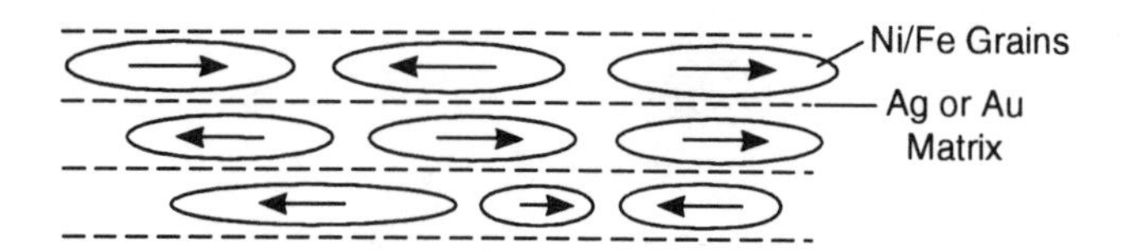

Fig. 12.10. A magnetization state of a permalloy–nonmagnetic metal two-phase GMR structure in its demagnetized (zero external field) state.

It is important to realize that both the superlattice and granular permalloy GMR materials can be made without the requirement for ultra-high-vacuum MBE machines. The fact that these materials can be made with ordinary laboratory sputtering equipment is surely an augur of the future in GMR device development.

Further Reading

Additional reading material is listed here that will prove helpful to readers who seek more detailed information.

Baibich, M. N., Broto, J. M., Fert, A., Nguyen Van Dau, F., and Petroff, F. (1988), "Giant Magnetoresistance of (001)Fe/(001)Cr Magnetic Superlattices," *Phys. Rev. Lett.* **61,** 21.

Dieny, B., Speriosu, V. S., Parkin, S. S. P., Gurney, B. A., Wilhoit, D. R., and Mauri, D. (1991), "Giant Magnetoresistance in Soft Ferromagnetic Multilayers," *Phys. Rev. B* **43,** 1.

Hylton, T. L., Coffey, K. R., Parker, M. A., and Howard, J. K. (1993), "Giant Magnetoresistance at Low Fields in Discontinuous NiFe-Ag Multilayer Thin Films," *Science,* p. 261.

White, R. L. (1992), "Giant Magnetoresistance—A Primer," *IEEE Trans.* **MAG-28,** 5.

CHAPTER

13

Spin Valve and Granular Giant Magneto-Resistive Heads

Although the giant magneto-resistive (GMR) effect has not yet been known to science for even 10 years, this has not deterred MRH designers from incorporating the effect into working prototype MRHs. Usually, the GMR structure or material has been used as the magneto-resistive element (MRE) in the shielded configuration discussed in Chapter 8. It is widely expected that GMR heads (GMRHs) will appear in mass production within a few years.

The attraction of using GMRHs lies both in their greater sensitivity and higher total change in resistance. In Chapter 11, the performance of an anisotropic shielded MRH was compared with that of an inductive head in terms of an equivalent *NV* product. For small-signal analog applications, the GMRH is superior in proportion to its low-field sensitivity, $d(\Delta\rho/\rho_0)dH$. In optimized large-signal digital GMRHs, their advantage is proportional to the magneto-resistive coefficient $\Delta\rho/\rho_0$.

Spin Valve GMREs

The structure of a typical spin valve GMRE is shown in Fig. 13.1. There are just two ferromagnetic layers of permalloy. They are called the *pinned* and the *free* layers. The pinned layer is exchange coupled to an underlying antiferromagnet with the requisite intimate atomic contact being indicated schematically in Fig. 13.1 by the interfacial zigzag line.

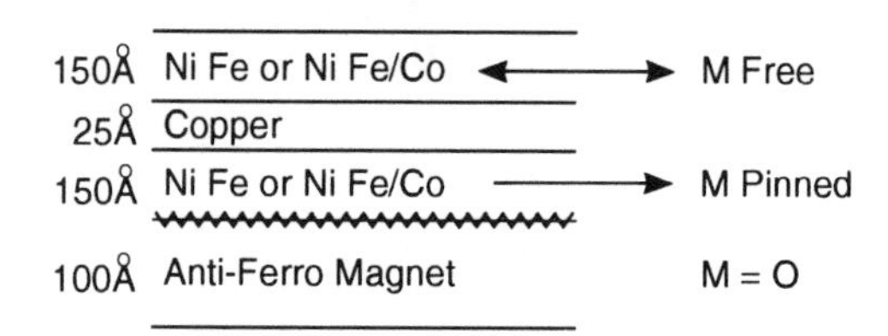

Fig. 13.1. A permalloy–copper spin valve GMR structure showing the exchange pinned and the free layers.

The atomic details of an idealized, epitaxial ferromagnet/antiferromagnet exchange-coupled interface is shown in Fig. 13.2. If every atom couples ferromagnetically across the interface, then very high exchange anisotropy can result. If, however, the atoms misregister by, say, one atom in a hundred, then the exchange coupling will be some average over the varying interatomic spacings and might be low. Alternately, if mono-atomic steps exist in the antiferromagnet, the overall coupling might again be low. Furthermore, if the antiferromagnet–ferromagnet interface is such that alternate spins in the antiferromagnet are oppositely oriented, which is the case in the so-called "compensation" crystallographic planes of some antiferromagnets, no coupling will occur. As has been pointed out in Chapter 5, the exact atomic details of the interface are critical and might not be measurable, because they are buried beneath the overlying permalloy layer.

The purpose of exchange coupling the pinned layer to the antiferomagnet is to establish its magnetization in a fixed direction. As discussed in Chapter 6, this direction can be set during the initialization processing of the head. Typically, the antiferromagnetic layer is a few hundred Å thick and is made of MnFe, CoO, or NiFeTb. The pinned layer is usually made of permalloy, for all the usual reasons. Because copper (but not silver) is soluble in permalloy, it is often over-coated

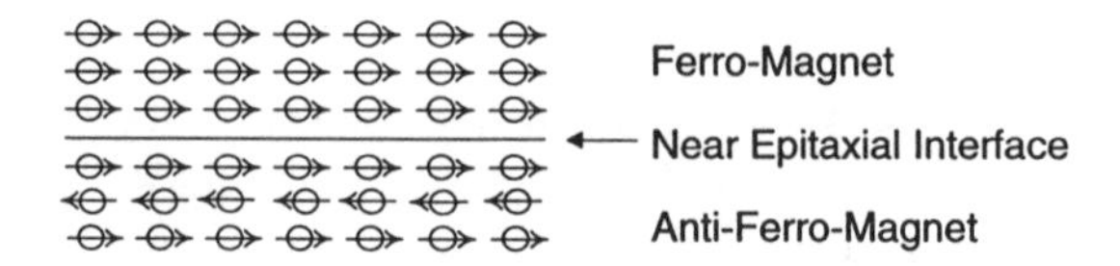

Fig. 13.2. The magnetic state of a nearly perfect, or epitaxial, interface between a ferromagnet and an antiferromagnet.

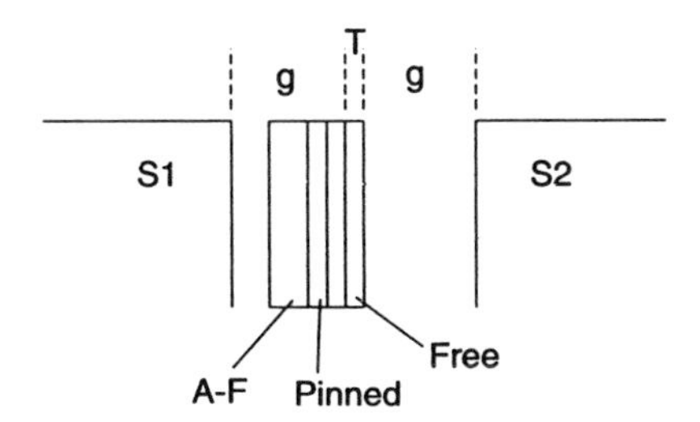

Fig. 13.3. The arrangement of a spin valve GMR structure in the gap of a shielded head.

("dusted") with a thin, 5- to 10-Å layer of cobalt, which acts as a diffusion barrier stopping the copper–permalloy interdiffusion.

Above the nonmagnetic layer, usually of copper, is the cobalt-dusted permalloy free layer. In the free layer, the anisotropy axis, defined during deposition or by magnetic annealing during initialization, is usually set to be orthogonal to the pinned magnetization direction. The reason for this "crossed" arrangement is discussed later.

Figure 13.3 shows how the entire spin valve structure is placed between shields to make a shielded GMRH. Note that ideally the free layer will be on the centerline between the shields so that self-biasing is avoided.

In Fig. 13.4, just the free and pinned layers are shown in perspective. The angle between their single-domain magnetization directions is β and the GMR is proportional to $-\Delta R/2 \cos \beta$.

The perspective diagram shown as Fig. 13.5 again shows the free and pinned layer magnetization directions and, additionally, the crossed easy axis of the free layer.

The equilibrium magnetization–easy axis angle θ, shown in Fig. 13.6,

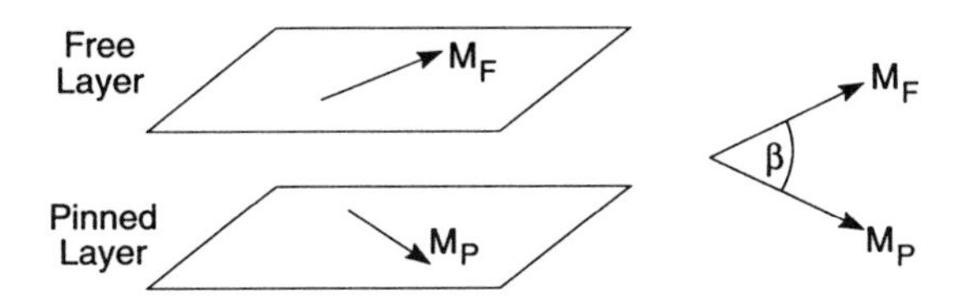

Fig. 13.4. The free and pinned layers of a spin valve showing the angle between the magnetizations.

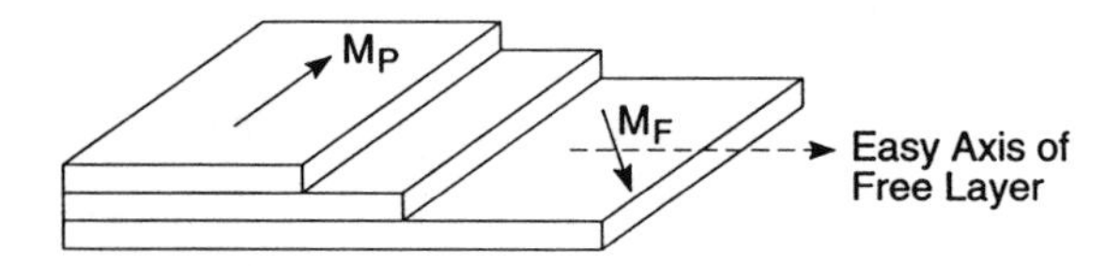

Fig. 13.5. A perspective diagram of a two-layer spin valve showing the easy axis of the free layer.

is, in the absence of demagnetizing field effects, simply given by $\sin\theta = H_y/H_k$ where H_y is the magnetic field from the recording medium and H_k is the anisotropy field. The angle between the free and pinned layer magnetizations, β, is equal to $(\pi/2 - \theta)$. Accordingly, the change in GMR, δR, is

$$\delta R = -\frac{\Delta R}{2}\cos\beta = -\frac{\Delta R}{2}\cos\left(\frac{\pi}{2} - \theta\right)$$

$$= -\frac{\Delta R}{2}\sin\theta = -\frac{\Delta R}{2}\frac{H_y}{H_k}. \tag{13.1}$$

The remarkable fact thus emerges that the change in resistance, δR, is directly proportional to the vertical field, or flux, from the recording medium. This linear characteristic is shown in Fig. 13.7. Note that this intrinsically linear characteristic has been obtained without the use of any of the vertical bias techniques that are required in anisotropic MRHs. It is, or course, still necessary to provide appropriate horizontal bias in order to stabilize the edge domains.

Naturally, in a real device the finite dimensions of the GMRE cause nonuniform demagnetizing fields to exist. These make it increasingly

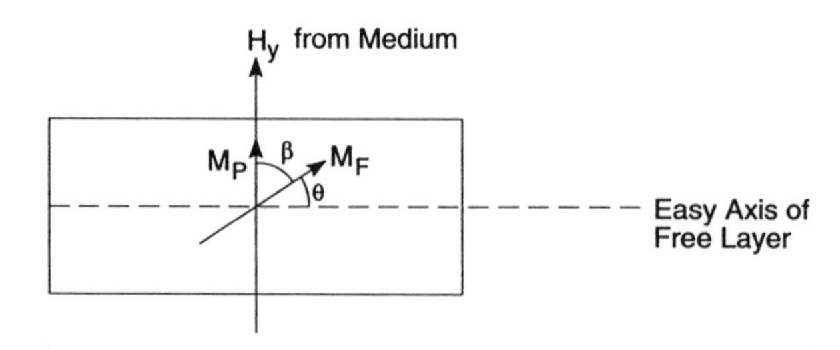

Fig. 13.6. The effect of the recording medium field on the free layer magnetization angle.

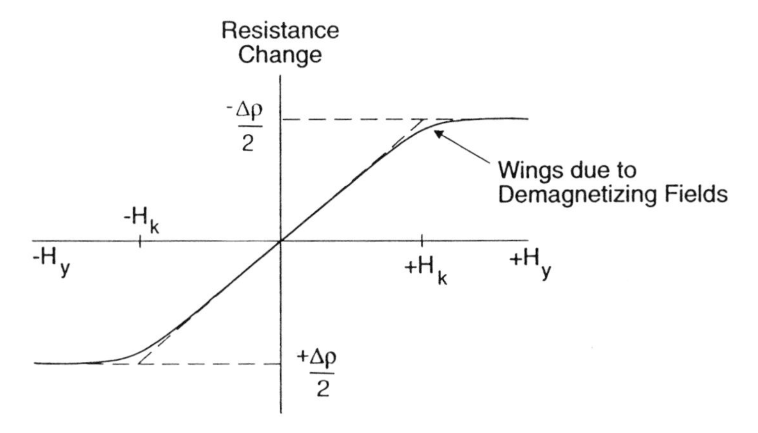

Fig. 13.7. The GMR resistance change versus field characteristic of a spin valve showing also the effect of nonuniform demagnetizing fields.

difficult to magnetize the free layer in the vertical (up or down the gap) directions. The effect of the nonuniform demagnetizating fields causes the resistance change characteristic to enter saturation gradually, displaying the wings shown in Fig. 13.7.

In practice, it is found that the spin valve resistance change characteristic is not of the exact odd symmetry shown in Fig. 13.7. The pinned layer is effectively a permanent magnet and it produces, by magnetostatic coupling, an unwanted vertical bias field on the free layer. This vertical bias field may be cancelled out by operating the spin valve at exactly the correct magnitude (and polarity) of current so that the current bias field is equal and opposite to the pinned layer bias field. Note that a correctly operating spin valve GMRH has neither pulse amplitude asymmetry nor, to first order, side reading asymmetry.

The dynamic range of magnetization angle change, set by the acceptable (odd only) harmonic distortion in a spin valve head, is approximately $\pm 60°$ about the quiescent $\theta = 0°$. The anisotropic MRH dynamic range is about $\pm 30°$ about the quiescent 45°.

Note, however, that fractional change in the resistance is the same. Thus, $\Delta R/2(\cos 30° - \cos 150°) = 0.87\Delta R$ for the GMR head, whereas $\Delta R(\cos^2 15° - \cos^2 75°) = 0.87\Delta R$ in the anisotropic MRH case.

Granular GMREs

The usual manner proposed for the use of granular GMR materials is to have the easy axis of the permalloy oriented in the track width W

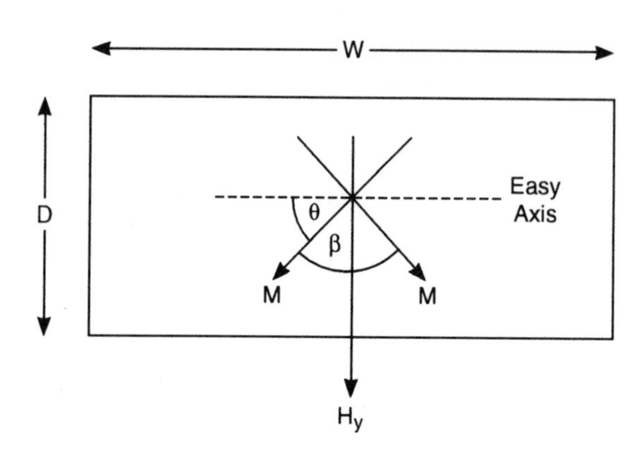

Fig. 13.8. The magnetization directions and the easy axis orientation for a granular GMRH.

direction as is shown in Fig. 13.8. Thus, when no field is applied, the structure is demagnetized ($M = 0$), with the single-domain permalloy grains magnetized at random in the two cross-track directions, and the material is in the high-resistance state. When, however, a vertical field H_y is applied, the magnetization in the grains rotates into parallelism and the material goes to the low-resistance state.

In the absence of demagnetizing fields, the resistance change characteristic can be deduced immediately. As shown in Fig. 13.8, the angle between the grain magnetizations M and the easy axis is θ, where $\sin\theta = H_y/H_k$. The angle between the two possible grain magnetization directions is β and it is equal to $(\pi - 2\theta)$. The change in resistance is

$$\delta R = -\frac{\Delta R}{2}\cos\beta = -\frac{\Delta R}{2}\cos(\pi - 2\theta) = \frac{\Delta R}{2}(1 - 2\sin^2\theta)$$

$$= \frac{\Delta R}{2}\left[\left(1 - \left(2\frac{H_y}{H_k}\right)^2\right]. \qquad (13.2)$$

This characteristic is shown as the dashed curve in Fig. 13.9. In reality, of course, nonuniform demagnetizing fields round out the discontinuous dashed curve shown in Fig. 13.9 and the actual granular GMRE characteristic is as shown by the solid line.

Note that this characteristic is not inherently linear as in the case of spin valve GMREs, but is in fact similar to that familiar in anisotropic

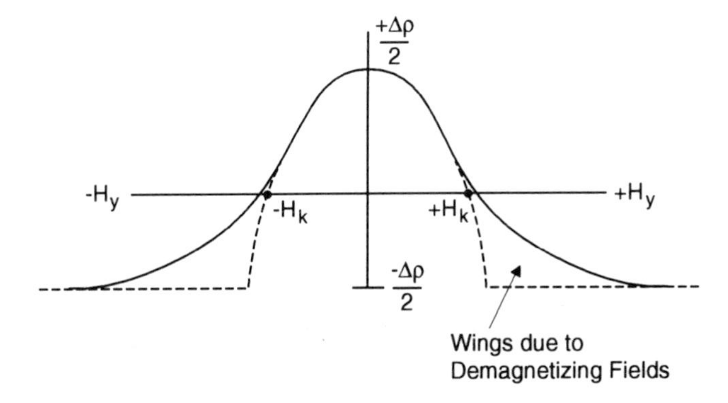

Fig. 13.9. The resistance change versus field characteristic of a granular GMR material showing also the effect of nonuniform demagnetizing fields.

MRHs. It follows that, in order to achieve satisfactory linear operation with granular MRHs, it is necessary to supply a vertical bias field. Additionally, horizontal biasing is required so that the end regions are kept in a fixed magnetic state.

Further Reading

Additional reading material is listed here that will prove helpful to readers who seek more detailed information.

Tsang, C., Fontana, R., Lin, T., Heim, D., Speriosu, V., Gurney, B., and Williams, M. (1994), "Design Fabication and Testing of Spin-Valve Read Heads for High Density Recording," *IEEE Trans.* **MAG-30,** 6.

CHAPTER

14

Simplified Design of a Shielded Magneto-Resistive Head

In this chapter, a simplified design procedure for a single-element, anisotropic magneto-resistive effect shielded head is demonstrated by means of a numerical example. This design procedure is included in this book because it shows three things very clearly. First, it shows precisely which physical phenomenon controls the major aspects of a shielded MRH design. Secondly, it demonstrates that in order to obtain optimum performance from a MRH it *must match properly* the magnetic transition in the recording medium. Third, it demonstrates how the chosen units work. Various different units are used in this chapter in accordance with the conventions used in the disk-drive industry in the United States.

The Written Magnetization Transition

The write-head geometry is shown in Fig. 14.1. The recording medium thickness, δ, is taken here to be 2 μin or 500 Å. The write head-to-recording medium flying length, d_w, is assumed to be 2 μin. or 500 Å. The recording medium is assumed to be similar to Co–Cr–Pt and have the following magnetic properties: saturation remanence (M_R) of 800 emu-cm^{-3} ($4\pi M_R \approx 10{,}000$ G) and coercivity (H_c) of 1600 Oe.

As was pointed out in Chapter 2, it is not necessary to specify the write-head gap length, g_W, because it is not an important parameter in the

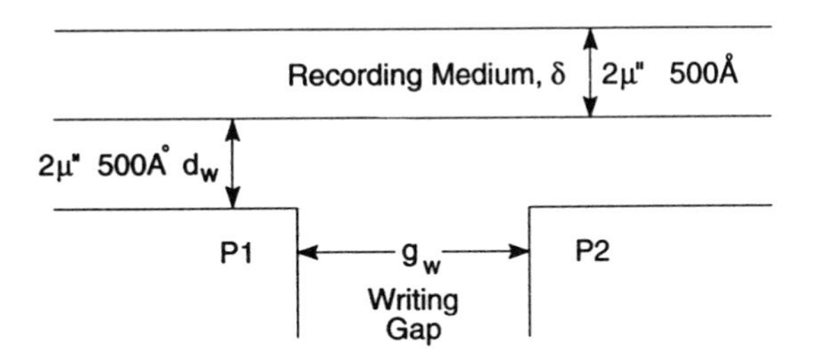

Fig. 14.1. The geometry of the recording medium/write-head interface with the critical dimensions shown.

writing process. Changes in the write-head gap length are, to a large extent, compensated for by adjusting the write current so that the deep gap field remains close to $3H_c$. In practice, the actual gaplength is set by the compromise between improving over-write (use a larger gap) and decreasing side-writing (use a smaller gap).

In the Williams and Comstock model discussed in Chapter 2, the written magnetization transition is assumed to have the arctangent form shown in Fig. 14.2. The remanent magnetization changes from $-M_R$ to $+M_R$ over a distance of πf, where f is the slope parameter and

$$f = 2\left[\frac{2}{\sqrt{3}}\frac{M_R}{H_c}\,\delta\left(d_w + \frac{\delta}{2}\right)\right]^{\frac{1}{2}}. \qquad (14.1)$$

Substituting $d_w = \delta = 2$ µin., $M_R = 800$ emu-cm^{-3}, and $H_c = 1600$ Oe yields $f = 3.72$ µin.

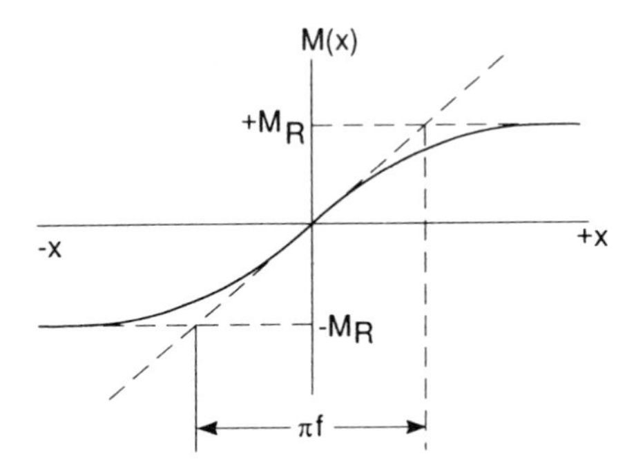

Fig. 14.2. The arctangent magnetization transition.

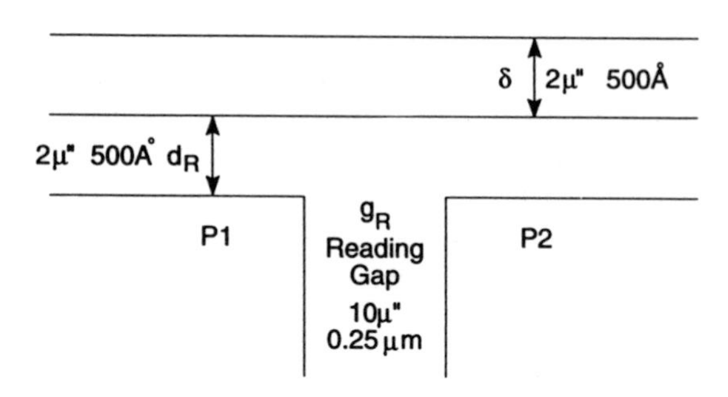

Fig. 14.3. The geometry of the inductive reading head/recording medium interface with the dimensions shown.

Inductive and Shielded MRH Output Pulses

An inductive head of gap length g_R is shown in Fig. 14.3. The head-to-medium flying height is d_R. This inductive head produces the output pulse shown in Fig. 14.4, which can be characterized by its 50% (of peak) amplitude width, PW_{50}. According to Chapter 3,

$$_{\mathrm{IND}}\mathrm{PW}_{50} = 2[(d_R + f)(d_R + \delta + f) + g_R^2/4]^{\frac{1}{2}}. \qquad (14.2)$$

In contrast to the write-head gap length, g_W, the read-head gap length, g_R, is one of the most critical parameters in system design. A gap length that is too narrow leads to heads of low efficiency. Gap lengths that are too wide produce undesirably wide output pulses. Here, we assume that a 25% increase in PW_{50} is acceptable. Substituting $f = 3.72$ μin., $\delta = d_R = 2$ μin., the pulse widths, $_{\mathrm{IND}}\mathrm{PW}_{50}$, for gap lengths of 0, 5, 10, and 15 μin. are 13.2, 14.2, 16.6, and 20 μin., respectively. Accordingly, an inductive read-head gap length g_R of 10 μin. or 0.25 μm is chosen as a satisfactory compromise.

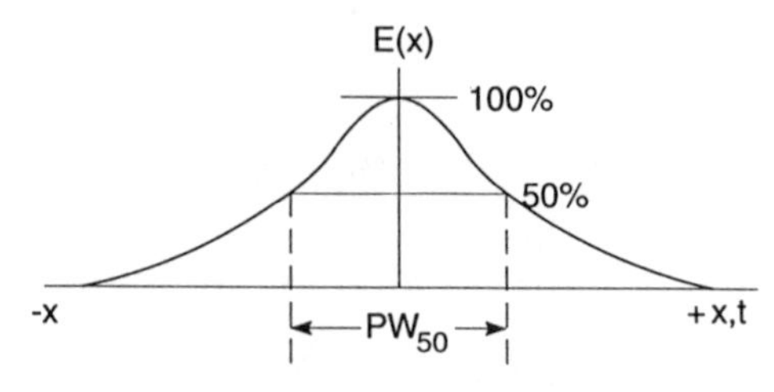

Fig. 14.4. The isolated magnetic transition output voltage pulse shape.

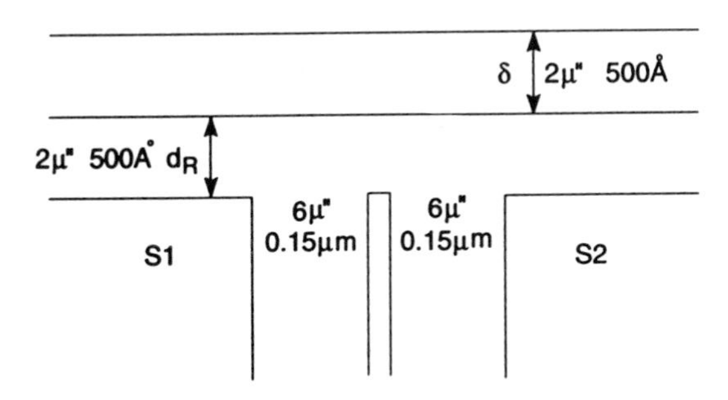

Fig. 14.5. The geometry of the shielded MRH recording medium interface with the dimensions shown.

Simple data detection systems, such as the venerable differentiating peak detector, are operated at minimum transition intervals about equal to PW_{50} yielding, in this case, a maximum flux reversal density of $(PW_{50})^{-1}$ or 60,000 frpi. With a more advanced postequalization and data recovery technique, such as PRML, a 50% increase to 90,000 frpi is feasible. When a channel code such as 2/3 rate (1,7) is used, the bit density is 4/3 the flux reversal density and, accordingly, 90,000 frpi corresponds to 120,000 bpi.

In Chapter 8, it was noted that an inductive head with a whole gap length of $1.6g$ has the same spatial resolution as does a shielded MRH with half-gap lengths of g. Thus, a shielded MRH with a half-gap g equal to $10/1.6 \approx 6.0$ μin. produces the same PW_{50} as the 10-μin. gap length inductive head that was selected earlier.

The half-gap length of 6 μin. or 0.15 μm shown in Fig. 14.5 is the first of the shielded MRH design parameters to be fixed and it should be

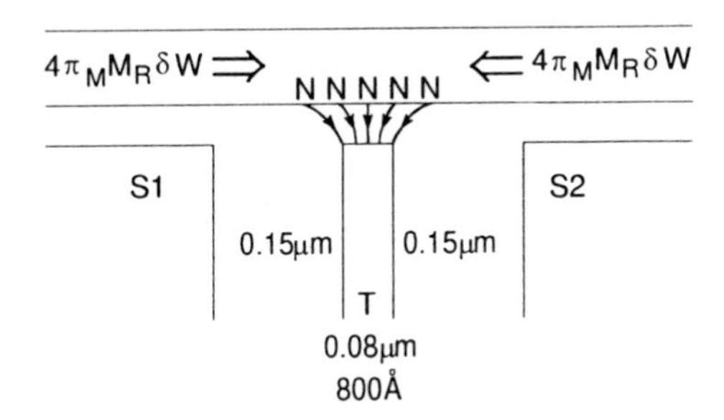

Fig. 14.6. The flux flow into the top of the shielded MRE from a magnetic transition directly over the gap.

noted that it is *intimately* related to the written transition. The MRE thickness T and depth D remain to be determined.

The MRE Thickness T

To determine the MRE thickness T, we must make sure that the top of the element closest to the recording medium remains unsaturated as the magnetization transition passes over it. This situation is shown in Fig. 14.6.

The flux change in the recording medium is $8\pi M_R\delta W$ because, regardless of the details of the magnetization transition, the flux in the medium is $\pm 4\pi M_R \delta W$ at positions that are far away from the transition. It is assumed that all of this flux change has to be absorbed by the MRE.

At the top of the MRE element, the magnetization angle θ is close to zero. Despite the fact that the vertical bias has established an average angle that is close to 45°, demagnetizing fields at the edges force the magnetization to be almost parallel to the edge. The maximum flux capacity at the top of the MRE is, therefore, approximately $B_s TW$.

Equating these fluxes gives the minimum MRE thickness $T = 8\pi M_R \delta / B_s$. Putting B_s of the permalloy equal to 12,500 G, the minimum thickness is 1.6δ. To avoid saturation in the MRE, the single-element permalloy MRE must, therefore, be at least 3.2 μin. or 800 Å thick.

When the design includes another highly permeable element, such as a soft adjacent layer (SAL), the MRE must be proportionately thinner. If the SAL and MRE saturation inductions and thicknesses are B_s, B_M, T_s, and T_M, respectively, the flux saturation criterion becomes $B_s T_s + B_M T_M \approx 8\pi M_R \delta$. If, as is frequently the case, the flux capacity $B_s T_s$ of the SAL is almost the same as that of the MRE, the minimum MRE thickness in the numerical design becomes 1.6 μin. or 400 Å.

The MRE Depth D

The MRE element depth is determined by the requirement that the depth D be less than l, the magnetic transmission line depth discussed in Chapter 8. If the element is significantly deeper than l, the maximum flux efficiency (≈50%) of the shielded MRH design cannot be realized. If the dimension D is too small, there is the possibility that the design will be needlessly difficult to fabricate with high process yields.

Transmission line length $l = [Tg\mu/2]^{\frac{1}{2}}$, where μ is the low-field permeability of the MRE. The permeability of permalloy can be taken to be approximately equal to B_s/H_k ($\approx$1600). With T = 3.2 μin., g = 6 μin., l = 124 μin. or 3.1 μm. For ease of fabrication, D = 3 μm is chosen here.

When another permeable element is present, such as a SAL (or another permalloy layer in the case of a double-element MRH), note that, provided it has similar properties to the MRE, the transmission line length is not changed significantly.

The existence in the gap of two similar permeable elements, each of one-half thickness, does not change appreciably the flux side leakage because very little flux flows from one element to the other. This situation is shown in Fig. 14.7, where the half-fluxes decrease linearly down the depth D of 3 μm.

Now all the principal dimensions, g, T, and D, of the simplified shielded MRH design are known.

The Measuring Current *I*

It is, of course, desirable to operate the MRH with the highest possible measuring current, I, because the output signal amplitude is directly proportional to I. Two different physical phenomena, however, limit the current magnitude that can be used.

The first operational limit is the I^2R joule heating of the MRE. This heat has to be conducted away by the structure surrounding the MRE. Generally, MREs are operated at temperatures of about 30°C above ambient. If the current were then to be doubled, the temperature rise would then become four times greater (120°C). Joule heating causes the MRE

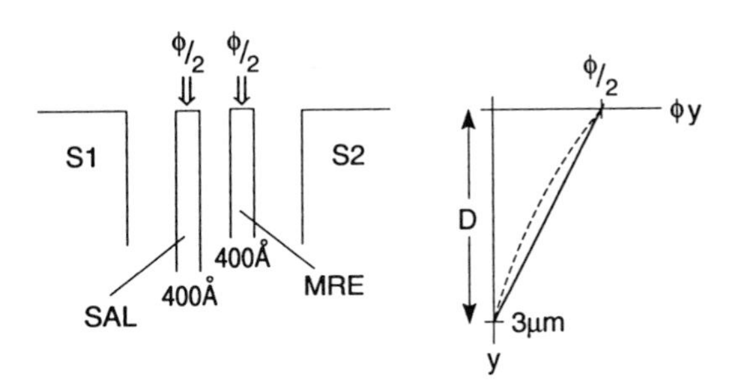

Fig. 14.7. The flux flow in a SAL-biased MRE, showing the division of the flux into the two permeable layers.

temperature to rise very quickly since most MRHs have thermal diffusion time constants of a few tens of microseconds.

The second limit is set by the phenomenon of electromigration. In electromigration, the repeated collisions of the conduction electrons against the scattering centers eventually cause bulk movement or migration of the conductor material. In permalloy, serious electromigration occurs whenever the current density, J_0, is higher than 10^7 A-cm^{-2}. The critical current density is higher in materials where the atoms are more tightly bound and, thus, have higher melting temperatures. Examples of such conductors are Pt, Ta, W, and other so-called "refractory" metals.

The electromigration phenomenon is invidious because the material transfer depends not only on the current density J_0, but also on the accumulated time in service of the MRE. Whereas the joule heating limits can be established in seconds, the determination of the electromigration limit can take many thousands of hours of device testing.

Assuming that the allowable current is set by electromigration concerns, the measuring current is J_0TD. Substituting $J_0 = 10^7$ A-cm^{-2}, $T = 800$ Å or 8×10^{-6} cms, and $D = 3$ μm or 3×10^{-4} cms gives $I =$ 24 mA.

The Isolated Transition Peak Output Voltage δV

The specific peak output voltage of an *optimally designed* permalloy MRH has already been discussed in Chapter 11. It is about 2 V per centimeter of track width W. For convenience, however, the basic arguments leading to this result are repeated here.

Of the total magneto-resistive change (4% in permalloy) only one-half is available for negative and positive pulses. The shielded MRH has an efficiency of one-half when $D < l$. The resistance $R = \rho_0 W/TD$. The measuring current $I = J_0TD$. The peak output voltage is, therefore, repeating Eq. (11.6),

$$\delta V = \frac{1}{2}\,(10^7\,TD)\left(\frac{2}{100}\right)\,(20 \cdot 10^{-6})\,\frac{W}{TD} = 2W \qquad \text{[volts]}. \qquad (14.3)$$

Again, note that when electromigration sets the measuring current density limit, the peak output voltage is independent of the dimensions T and D.

Alternative designs were considered earlier in this chapter, where there are two permeable elements, such as the SAL and the MRE, in the gap. Clearly, the specific peak output voltage is *exactly the same* in these cases, provided the proper changes in MRE thickness and measuring current are made.

The maximum possible output voltage of a permalloy MRH is determined by the *intrinsic properties of the permalloy* itself and not by the details of the optimum design.

Now the simplified design is complete. For a 10,000 G remanence, 1600 Oe coercivity medium of 500 Å thickness, the peak pulse output voltage is 2 V per centimeter of track width at a measuring current of 24 mA, when the shielded MRH has half-gaps of 6 μin. and the MRE dimensions are 800 Å thick and 3 μm deep. All that remains now is a comparison of the shielded MRH peak output voltage with that of a comparable inductive head.

Optimized MRH versus Inductive Head Voltages

There are, of course, many ways in which the peak pulse output voltage of an inductive head can be determined. When the isolated pulse width PW_{50} is known, however, the method described next is both the simplest and most direct.

Consider the output pulse, shown in Fig. 14.8, with peak voltage, E_{peak} and 50% amplitude width, PW_{50}. The area, A, under the pulse may be approximated as $\frac{1}{2}E_{peak} \times 2PW_{50} = E_{peak} \times PW_{50}$. This area is, of course, proportional to the flux change $\Delta\Phi = 8\pi M_R \delta W$ of the digital transition because

$$E(x) = -10^{-8}\, N\, d\phi/dt = -10^{-8}\, NV\, d\phi/dx \qquad \text{[volts]} \quad (14.4)$$

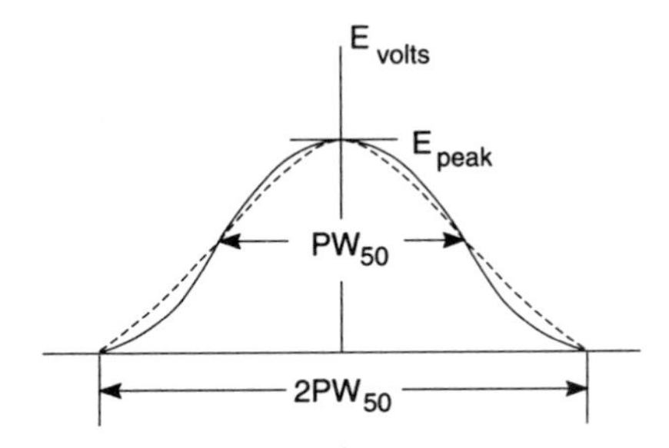

Fig. 14.8. The output voltage pulse of an inductive head and its triangular approximation.

so that

$$A = \int E(x)\,dx = -10^{-8}\,NV \int d\phi/dx\,dx = -10^{-8}\,NV\,\Delta\phi. \tag{14.5}$$

It then follows that

$$E_{\text{peak}} = 10^{-8}\,\frac{NV\,(8\pi M_R \delta W)}{\text{PW}_{50}}. \tag{14.6}$$

Upon substituting $M_R = 800$ emu-cm^{-3}, $\delta = 5 \times 10^{-6}$ cms, and $\text{PW}_{50} = 16.6$ μin. or 42×10^{-6} cms, the result is $E_{\text{peak}} = 2.4 \times 10^{-5}\,NVW$ volts.

It follows that the maximum peak pulse of the optimized MRH ($2W$ volts) is equal to that of an inductive head with an NV product equal to 83,333 cms-sec^{-1}. In Chapter 11, equivalence was assumed to occur with $NV = 80{,}000$ cm sec^{-1} solely in the interest of numerical simplicity.

It should be realized that the actual value of the equivalent NV product depends upon the details of the head–medium interface and is not to be regarded as an invariable constant. In the next section, the behavior of a so-called "MR disc"–shielded MRH interface is summarized and it is shown to have an equivalent NV product over double that calculated above.

"MR Disc"–Shielded MRH Performance

An obvious design trend in rigid disc technology is the introduction of the so-called "MR disc." These discs exploit the fact that an *optimally designed* MRH, using a permalloy MRE, will always produce the same output voltage (2 volts per centimeter of trackwidth). Accordingly, the $M_R\delta$ product of the disk may be reduced in order to realize both a smaller write process slope parameter, f, and pulse width, PW_{50}. Reductions in $M_R\delta$ are achieved by using thinner films and also by reducing the Co content of the CoCrPt alloy.

Very recent (1995) "MR discs," with a $Co_{60}\,(CrPt)_{40}$ composition ($M_R \approx 400$ emu/cm^2) and $\delta = 1$ μ" (250 Å) have an $M_R\delta$ product of 1 milli-emu/cm^2 only. In comparison, the disc considered in the numerical design example worked out above has $M_R\delta = 4$ milli-emu/cm^2.

The reader may care to verify that the following set of parameters is numerically self-consistent:

Remanent magnetization, M_R	400 emu/cm^2
Coercivity, H_c	1,600 Oe
Film thickness, δ	250 Å
Writing spacing, d_w	500 Å
Slope parameter, f	425 Å
Reading spacing, d_R	500 Å
S-MRH half-gap length, g	830 Å
Pulse-width, PW_{50}	10 microinches
MRE thickness, T	200 Å
Flux decay length, l	1.15 micron
Measuring current, I	2.3 mA (at $J_0 = 10^7$ A cm^{-2})
Peak output voltage	2 volt/cm (4%), 1 volt/cm (2%)
Equivalent NV product	200,000 cm sec (4%)
	100,000 cm/sec (2%)

It should be noted, carefully, that the most important result of using the 1 milli-emu/cm^2 "MR disc" is, from the recording system point of view, the substantial reduction the slope parameter, f, and the pulse width, PW_{50}. As was discussed in Chapter 2, the linearized transition width is πf or, in this case, 1335 Å. The pulse width is now 10 μin or only 66% that calculated for the 4 milli-emu/cm^2 disc. Such a narrow pulse width would permit about 100,000 frpi with simple detectors, about 150,000 frpi with advanced detectors and about 200,000 bpi with advanced detectors and 2/3 rate (1,7) RLL coding.

The lower limit to useful reductions in $M_R\delta$ is set by the decreasing magneto-resistive coefficient, $\Delta\rho/\rho_0$, associated with the lower MRE thickness, T, required.

Further Reading

Additional reading material is listed here that will prove helpful to readers who seek more detailed information.

Tsang, C., Chen, M. M., Yogi, T., and Ju, K. (1990), "Gigabit Density Recording Using Dual-element MR/Inductive Heads on Thin-film Discs," *IEEE Trans.* **MAG-26,** 5.

CHAPTER

15

Read Amplifiers and Signal-to-Noise Ratios

Magneto-resistive heads (MRHs) can be operated with two types of read amplifier. Although the more common is the high input impedance, voltage-sensing type, it is possible to use a low input impedance current-sensing design. Some of the relative advantages of these two approaches are discussed in this chapter.

Magneto-resistive heads are used because they provide important advantages over inductive reading heads. Two of these advantages, namely, the high output voltage and its independence of MRH-to-recording medium relative velocity, have already been discussed. In this chapter, the principal factors controlling the signal-to-noise ratio (SNR) of a MRH are reviewed.

Magneto-resistive heads are expected to facilitate the attainment of extremely high areal densities ($\approx$10 Gbits/in.^{2}) in the future. This chapter concludes with a short discussion of the specific characteristics of MRHs that encourage these predictions.

MRH Read Amplifiers

The electrical circuit model of an MRH is shown in Fig. 15.1. The fixed part of the resistance of the MRE, R_0, and the resistance, R, and inductance, L, of the connecting leads are shunted by the stray capaci-

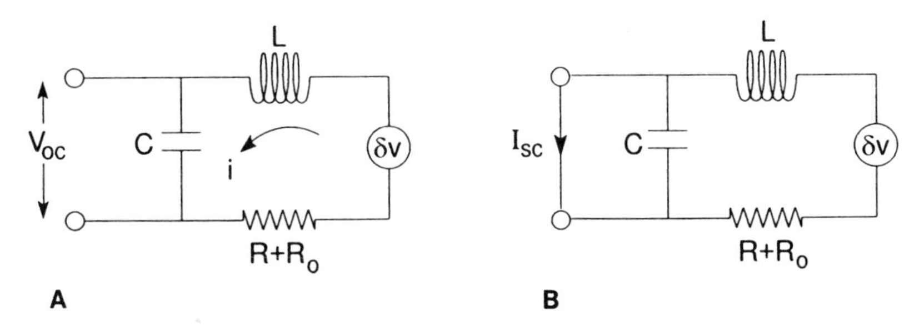

Fig. 15.1. Electrical circuit model of a MRH for (A) voltage sensing and (B) current sensing.

tance, C, of the leads. The magneto-resistive effect (MRE) is represented by a voltage generator, $\delta V = I\delta R$, where I is the measuring current and δR is the variable part of the resistance of the MRE. In Fig. 15.1, the measuring current I is not depicted.

When a high internal impedance, read amplifier is connected to the head terminals, it measures the open-circuit voltage, V_{oc}, as shown in Fig. 15.1(A). The signal current, i, flowing in the loop is

$$i = \frac{\delta V}{(R + R_0) + j\omega L + 1/j\omega C}, \tag{15.1}$$

and the output voltage is

$$V_{0c} = \frac{i}{j\omega C} = \frac{\delta V}{j\omega(R + R_0)C + (1 - \omega^2 LC)}. \tag{15.2}$$

The frequency response is second order, and has a resonant frequency of $\omega = 1/\sqrt{LC}$. The resonance is undesirable because it causes an oscillating or ringing output voltage. Very similar considerations arise, of course, in voltage sensing with inductive read heads. However, in MRHs the inductance (typically 20 nH) is very much lower than in inductive heads (typically 1000 nH) and it follows that the resonant frequency is higher by a factor of approximately $\sqrt{50} = 7$. The low inductance of MRHs is due to their "single-turn" structure and it is another of their many advantages.

A power supply with a high internal impedance is a constant current source. Accordingly, high impedance read amplifier is used to produce a fixed resistance change measuring current I. This mode of operation is often called the current-forcing, voltage-sensing mode.

The other mode of operation uses a read amplifier that has a low internal impedance and it is termed the voltage-forcing, current-sensing mode. The low impedance of the amplifier has the important effect of effectively shunting out the stray capacitance C.

The low impedance amplifier senses the short circuit current, I_{sc}, shown in Fig. 15.1(B), where

$$I_{sc} = \frac{\delta V}{(R + R_0) + j\omega L}. \tag{15.3}$$

The frequency response is, therefore, first order having no undesirable resonance. Instead, the spectrum merely rolls off with a -3 dB frequency $\omega = (R + R_0)/L$. An advantage of this design is the improved high frequency response.

Current-sensing amplifiers are widely used with inductive heads in digital video recorders, where the operating frequencies (50 to 100 MHz) are higher than in computer peripheral recorders. When current sensing is used with inductive heads, the amplifier essentially acts as an integrator and the output follows $M(x)$ rather than the usual $dM(x)/dx$. Integration, which is multiplication by $1/j\omega$ in the frequency domain, occurs because the impedance of the inductive head is mainly inductive, so that $I_{sc} \approx \delta V/j\omega L$. In the case of MRHs, however, the head impedance is principally resistive and $I_{sc} \approx \delta V/(R + R_0)$ and integration does not occur.

Because the MRE is carrying the measuring current I, provision has to be made to control electrical shorting or arcing whenever it touches a conductive recording medium such as a thin-film disk. In some read amplifier designs, the electrical potential of the MRE is held constant, by a servo circuit, so that it is equal to that of the disk. The disk potential can either be ground or some other value. When the disk is "floated," that is, held at a potential other than ground, electrical insulation of the disks, spindle bearings, or motor is, of course, necessary. In "single-ended" read amplifiers, one end of the MRE is simply held at ground potential and disk floating is not needed.

It is obvious from the preceding discussion that there are many different designs for MRH read amplifiers. Which design will be most generally adopted is a question for the future.

Signal-to-MRH Noise Ratios

Throughout this book it has been assumed implicitly that the performance of a MRH is determined by the magneto-resistive coefficient, $\Delta\rho/\rho_0$. In reality, however, this need not always be the case.

Consider the situation where electromigration limits the allowable current density, J_0. In permalloy, $J_0 \approx 10^7$ A-cm^{-2}, but, if other materials are considered for MREs, no doubt their J_0 values will differ. The signal voltage, $\delta V = I\delta R$, is simply proportional to $\Delta\rho$, the maximum MR resistance change. If the MRE resistance is R, the mean thermal noise voltage generated is $(4kTR\Delta f)^{\frac{1}{2}}$ and it is proportional to $\sqrt{\rho_0}$. Here, k is Boltzmann's constant, T is the absolute temperature in °K, and Δf is the bandwidth in hertz. It follows that the MRH-limited SNR is proportional to $\Delta\rho/\sqrt{\rho}$ and not the usual $\Delta\rho/\rho_0$.

On the other hand, suppose that in some other MRE material it is joule heating (I^2R) that limits the measuring current, so that I is proportional to $1/\sqrt{R} = 1/\sqrt{\rho_0}$. In this case, the signal voltage $\delta V = I\delta R$ becomes proportional to $\Delta\rho/\sqrt{\rho_0}$ and the MRH-limited SNR is now proportional to $\Delta\rho/\rho_0$.

Consider now the MRH-limited SNR of the optimized shielded MRH that was designed in Chapter 14. The peak signal voltage is $2W$ volts. The thermal noise voltage is $(4kTR\Delta f)^{\frac{1}{2}}$ where the resistance is $\rho_0\, W/TD$ ohms. Substituting $\rho_0 = 20 \times 10^{-6}\ \Omega$-cm, and the values $T = 400$ Å and $D = 3$ μm appropriate to the SAL design, the resistance $R = 1.6 \times 10^4 W\ \Omega$. Assuming Δf, the system bandwidth is 10 MHz, the thermal noise is approximately equal to $5 \times 10^{-5}\sqrt{W}$ V rms. The MRH-limited SNR (peak to rms) of the optimized MRH is, therefore, approximately $40{,}000\sqrt{W}$.

Note that both the MRH noise and the recording medium noise-limited SNR's depend on $\sqrt{W}$. It follows that the ratio of the two SNRs is independent of track width. This is one of the most important reasons for the optimism that MRHs will facilitate extremely high areal density magnetic recording.

Suppose that the optimized shielded MRH of Chapter 14 is used at a track width of $W = 5$ μm. The head-limited SNR is $40{,}000\ (5 \times 10^{-4})^{\frac{1}{2}} \approx 900$ or 59 dB. This value is much greater than the medium-limited SNR, which may be as low as 20 dB. The conclusion is not only that the recording system is recording medium noise-dominated but also that it will remain so at other track widths W.

Another way of thinking about the low noise of MRHs is to ask "At what track width does the MRH-limited SNR fall to the value 30 or 30 dB?" Solving 40,000 $\sqrt{W} = 30$ yields $W = 56 \times 10^{-8}$ cm (56 Å). Such a track width is, of course, far below the limits of the photolithographical processes used in making thin-film and MR heads.

It is safe to conclude that an optimally designed shielded MRH generates so little thermal noise that recording systems can remain medium-noise-limited at all accessible track widths. Moreover, when the promise of permalloy-based giant MRHs is realized, both the optimized output voltages and the head-limited SNRs will increase in proportion to the increase in ΔR. Output levels of $10W$ volts/cm (1 kV/m) can be anticipated in giant shielded MRHs.

Conclusions

Magneto-resistive heads, after early and unsuccessful attempts in analog recording, seem to be likely to find applications in every kind of digital recorder. There is widespread optimism that both anisotropic MRHs and giant MRHs will make possible areal densities approaching 10 Gbits/in.^{2} in both tapes and hard disks.

To understand why this revolution in reading heads is occurring, it is this writer's opinion that the reader can do no better than to review, once again, the advantages of MRHs as propounded by Hunt in 1970. The first paragraph of Hunt's paper, given in Chapter 7, stands as a model of clarity, brevity, and scientific accuracy.

Further Reading

Additional reading material is listed here that will prove helpful to readers who seek more detailed information.

Grochowski, E., and Thompson, D. (1994), "Outlook for Maintaining Areal Growth Rate in Magnetic Recording," *IEEE Trans.* **MAG-30,** 6.

Klaassen, K. B., and van Pedden, J. C. L. (1995), "Read/Write Amplifier Design Considerations for MR Heads," *IEEE Trans.* **MAG-31,** 2.

Mallinson, J. C. (1969), "Maximum Signal-to-Noise Ratio of a Tape Recorder," *IEEE Trans.* **MAG-5,** 3 (also in R. M. White; see Bibliography).

Mallinson, J. C. (1991), "A New Theory of Recording Media Noise," *IEEE Trans.* **MAG-27,** 4.

Murdock, E. S., Simmons, R. F., and Davidson, R. (1992), "Roadmap to 10 Gbt/in^2 Media," *IEEE Trans.* **MAG-28,** 5.

Appendix

CGS-EMU and MKS-SI (Rationalized) Units

This appendix consists of two tables. Table A.1 is a conversion table provided for the immediate conversion of the cgs-emu units used in this book into the corresponding MKS-SI units.

TABLE A.1
CGS-EMU to MKS-SI Units Conversion Table

Quantity	Symbol	cgs-emu	Conversion factor, C	MKS-SI
Magnetic flux density	B	Gauss	10^{-4}	tesla
Magnetic flux	ϕ	maxwell	10^{-8}	weber
Magnetomotive force	mmf	gilbert	$10/4\pi$	ampere
Magnetic field	H	oersted	$10^3/4\pi$	A/m
Magnetization	M	emu/cm^3	10^3	A/m
Magnetization	4πM	gauss	$10^3/4\pi$	A/m
Specific magnetization	σ	emu/g	1	A · m^2/kg
Magnetic moment	μ	emu	10^{-3}	A · m^2
Susceptibility	χ	Dimensionless	4π	Dimensionless
Permeability	μ	Dimensionless	$4\pi \times 10^{-7}$	H/m
Demagnetization factor	N	Dimensionless	$1/4\ \pi$	Dimensionless

TABLE A.2
See Where the 4π Factor Goes!

	MKS-SI (rationalized)	cgs-emu
Field of a wire	$H = \frac{2I}{(4\pi)R}$, A/m	$H = \frac{0.2I}{R}$, oersted
Field of a solenoid	$H = \frac{NI}{l}$, A/m	$H = \frac{0.1(4\pi)NI}{l}$, oersted
mmf	mmf $= NI$, A	mmf $= 0.1(4\pi)NI$, gilbert
Magnetic moment	$\mu = \int H\,dv$, A · m²	$\mu = \frac{1}{(4\pi)} \int H\,dv$, emu
Magnetization	$M = \frac{1}{VOL}\,\Sigma\mu$, A/m	$M = \frac{1}{VOL}\,\Sigma\mu$, emu/cm²
Flux density	$B = \mu_0(H + M)$, tesla	$B = H + (4\pi)M$, gauss
Permeability of free space	$\mu_0 = (4\pi)10^{-7}$, number	$\mu_0 = B/H = 1$, number
Flux	$\phi = \int B\,dA$, weber	$\phi = \int B\,dA$ maxwell
Voltage	$E = -N\,\frac{d\phi}{dt}$, V	$E = -10^{-8}\,N\,\frac{d\phi}{dt}$, V

The second table will be of interest to those readers with greater curiosity. Table A.2 gives the defining equations of all the magnetic entities used in this book, for both the cgs-emu and the MKS-SI (rationalized) systems of units. Notice that the difference between the two systems is not only a mere substitution of meters for centimeters and kilograms for grams, but also it includes a purely arbitrary shuffling of the position of the inevitable factor 4π. Accordingly, the second table has been given the title "See Where the 4π Factor Goes!" In both sets of equations, current and voltage are amperes and volts, respectively.

Further Reading

Additional reading material is listed here that will prove helpful to readers who seek more detailed information.

Brown, W. F. (1984), "Tutorial Paper on Dimensions and Units," *IEEE Trans.* **MAG-20,** 1.

Recommended Bibliography on Magnetic Recording

Bertram, H. Neal, *The Theory of Magnetic Recording,* Cambridge University Press, Cambridge, 1994. A highly mathematical and difficult-to-read treatment of the writing, reading, and noise processes in magnetic recording.

Mallinson, John C., *The Foundations of Magnetic Recording,* Academic Press, San Diego, 1993. A nonmathematical, but scientifically accurate, overview of most of the fundamental ideas used in magnetic recording media, heads and systems.

Mee, C. Denis, and Daniel, Eric D., Eds., *Magnetic Recording Handbook,* McGraw-Hill, New York, 1995. A comprehensive collection of contributions, by some 20 experts, on most of the important technologies used in and applications of magnetic recording systems.

Ruigrok, Jaap J. M., *Short Wavelength Magnetic Recording,* Elsevier Advanced Technology, Netherlands, 1990. A highly detailed and easy-to-read survey of both the theory and the practice of high-density magnetic recording.

White, R. M., Ed., *Introduction to Magnetic Recording,* IEEE Press, New York, 1985. A most useful collection of reprints of original research papers.

Index